INTERNATIONAL UNION OF CRYSTALLOGRAPHY
BOOK SERIES

IUCr BOOK SERIES COMMITTEE

Ch. Baerlocher, *Switzerland*
G. Chapuis, *Switzerland*
P. Colman, *Australia*
J. R. Helliwell, *UK*
K.A. Kantardjieff, *USA*
T. Mak, *China*
P. Müller, *USA*
Y. Ohashi, *Japan*
A. Pietraszko, *Poland*
D. Viterbo (*Chairman*), *Italy*

IUCr Monographs on Crystallography

1. *Accurate molecular structures*
   A. Domenicano, I. Hargittai, editors
2. *P.P. Ewald and his dynamical theory of X-ray diffraction*
   D.W.J. Cruickshank, H.J. Juretschke, N. Kato, editors
3. *Electron diffraction techniques, Vol. 1*
   J.M. Cowley, editor
4. *Electron diffraction techniques, Vol. 2*
   J.M. Cowley, editor
5. *The Rietveld method*
   R.A. Young, editor
6. *Introduction to crystallographic statistics*
   U. Shmueli, G.H. Weiss
7. *Crystallographic instrumentation*
   L.A. Aslanov, G.V. Fetisov, J.A.K. Howard
8. *Direct phasing in crystallography*
   C. Giacovazzo
9. *The weak hydrogen bond*
   G.R. Desiraju, T. Steiner
10. *Defect and microstructure analysis by diffraction*
    R.L. Snyder, J. Fiala and H.J. Bunge
11. *Dynamical theory of X-ray diffraction*
    A. Authier
12. *The chemical bond in inorganic chemistry*
    I.D. Brown
13. *Structure determination from powder diffraction data*
    W.I.F. David, K. Shankland, L.B. McCusker, Ch. Baerlocher, editors
14. *Polymorphism in molecular crystals*
    J. Bernstein
15. *Crystallography of modular materials*
    G. Ferraris, E. Makovicky, S. Merlino
16. *Diffuse x-ray scattering and models of disorder*
    T.R. Welberry
17. *Crystallography of the polymethylene chain: an inquiry into the structure of waxes*
    D.L. Dorset

18 *Crystalline molecular complexes and compounds: structure and principles*
    F. H. Herbstein
19 *Molecular aggregation: structure analysis and molecular simulation of crystals and liquids*
    A. Gavezzotti
20 *Aperiodic crystals: from modulated phases to quasicrystals*
    T. Janssen, G. Chapuis, M. de Boissieu
21 *Incommensurate crystallography*
    S. van Smaalen
22 *Structural crystallography of inorganic oxysalts*
    S.V. Krivovichev
23 *The nature of the hydrogen bond: outline of a comprehensive hydrogen bond theory*
    G. Gilli, P. Gilli
24 *Macromolecular crystallization and crystal perfection*
    N.E. Chayen, J.R. Helliwell, E.H. Snell
25 *Neutron protein crystallography: hydrogen, protons, and hydration in bio-macromolecules*
    N. Niimura, A. Podjarny
26 *Intermetallics: structures, properties, and statistics*
    W. Steurer, J. Dshemuchadse
27 *The chemical bond in inorganic chemistry: The bond valence model, 2e*
    I.D. Brown

IUCr Texts on Crystallography

1   *The solid state*
    A. Guinier, R. Julien
4   *X-ray charge densities and chemical bonding*
    P. Coppens
8   *Crystal structure refinement: a crystallographer's guide to SHELXL*
    P. Müller, editor
9   *Theories and techniques of crystal structure determination*
    U. Shmueli
10 *Advanced structural inorganic chemistry*
    Wai-Kee Li, Gong-Du Zhou, Thomas Mak
11 *Diffuse scattering and defect structure simulations: a cook book using the program DISCUS*
    R.B. Neder, T. Proffen
13 *Crystal structure analysis: principles and practice, second edition*
    W. Clegg, editor
14 *Crystal structure analysis: a primer, third edition*
    J.P. Glusker, K.N. Trueblood
15 *Fundamentals of crystallography, third edition*
    C. Giacovazzo, editor
16 *Electron crystallography: electron microscopy and electron diffraction*
    X. Zou, S. Hovmöller, P. Oleynikov
17 *Symmetry in Crystallography: Understanding the International Tables*
    P.G. Radaelli
18 *Symmetry relationships between crystal structures: applications of crystallographic group theory in crystal chemistry*
    U. Müller
19 *Small angle X-ray and neutron scattering from solutions of biological macromolecules*
    D.I. Svergun, M.H.J. Koch, P.A. Timmins, R.P. May
20 *Phasing in crystallography: a modern perspective*
    C. Giacovazzo
21 *The basics of crystallography and diffraction, fourth edition*
    C. Hammond

Symmetry in Crystallography

Understanding the International Tables

Paolo G. Radaelli

Clarendon Laboratory, Department of Physics, Oxford University

Great Clarendon Street, Oxford, OX2 6DP,
United Kingdom

Oxford University Press is a department of the University of Oxford.
It furthers the University's objective of excellence in research, scholarship,
and education by publishing worldwide. Oxford is a registered trade mark of
Oxford University Press in the UK and in certain other countries

© Paolo G. Radaelli 2011

The moral rights of the author have been asserted

First published 2011
First published in paperback 2016

All rights reserved. No part of this publication may be reproduced, stored in
a retrieval system, or transmitted, in any form or by any means, without the
prior permission in writing of Oxford University Press, or as expressly permitted
by law, by licence or under terms agreed with the appropriate reprographics
rights organization. Enquiries concerning reproduction outside the scope of the
above should be sent to the Rights Department, Oxford University Press, at the
address above

You must not circulate this work in any other form
and you must impose this same condition on any acquirer

Published in the United States of America by Oxford University Press
198 Madison Avenue, New York, NY 10016, United States of America

British Library Cataloguing in Publication Data

Data available

Library of Congress Cataloging in Publication Data

Data available

ISBN 978–0–19–955065–4 (Hbk.)
ISBN 978–0–19–878921–5 (Pbk.)

Links to third party websites are provided by Oxford in good faith and
for information only. Oxford disclaims any responsibility for the materials
contained in any third party website referenced in this work.

To Jacopo, Sofia (she spotted the pattern on page 47!) and Anna.

Preface

The purpose of this book is to provide a basic introduction to crystallographic symmetry, up to the point where understanding and using the *International Tables for Crystallography* (hereafter referred to as ITC) becomes possible. The ITC are an essential tool for understanding the literature and carrying out original research in the subfields of solid-state physics, chemistry and structural biology dealing with crystalline materials. Since their first edition, published in two volumes in 1935 under the title *Internationale Tabellen zur Bestimmung von Kristallstrukturen* with Carl Hermann as editor (Hermann, 1935), the ITC have steadily grown into eight ponderous volumes, to become the true "bible" of crystallographers. Over the years, the ITC have become progressively more general, from the initial topic of crystal structure determination by X-ray diffraction towards the recently stated purpose of covering all the subjects that may be of interest for crystallographers. Of course, information about crystal symmetry is central to the ITC, but subjects such as the properties of radiations used in crystallography, the physical properties of crystals and the proper format for crystallographic software are also covered. The ITC are a collective enterprise, each volume having one or more editors and many different contributors. As a consequence, although a remarkable effort has been made to provide a consistent notation and clear explanations, the perception of what is important and how it is best explained can vary from one section to the next, generating a significant amount of duplication and making the approach to the ITC a daunting task for the non-initiated. Here, I have collected a subset of topics regarding crystal symmetry, which I believe are most useful for physicists, chemists and biologists who deal with crystalline materials. The concepts are introduced gradually, with an emphasis on graphical rather than algebraic notation. This may seem somewhat surprising, but my experience over the years has been that manipulating operator "graphs" together with a schematic representation of the crystal structures is the clearest way to visualize symmetry and symmetry reduction through phase transitions. Once the basic concepts are firmly established, the necessary generation of atomic coordinates and their transformations, according to the different conventions, can be easily accomplished by one of the many crystallographic software programs now available on the market or in the public domain.

The approach I follow differs somewhat from the conventional way of teaching crystallography, which starts by establishing coordinate systems to locate atoms and then develops algebraic relations between the coordinates of symmetry-equivalent atoms; for an excellent book on general crystallography, including some elements of symmetry, see Giacovazzo *et al.* (2002). Here, I begin by examining simple patterns, as found in architecture, art, graphic design, etc., and develop a graphical notation to describe their symmetry and to combine symmetry operators. A minimal but reasonably rigorous set of group theoretical concepts, such as composition and the "multiplication table", will be introduced to enable the manipulations of these symbols. As the dimensionality and complexity of the patterns increase from symmetry around a fixed point to quasi-1D (frieze patterns) to 2D (plane or "wallpaper" patterns), so will the set of graphical symbols we become familiar with. I will also introduce in a graphical way concepts such as site symmetry, multiplicity and special positions (denoted using Wyckoff letters), group–subgroup relations, etc.

Only at this point will I introduce coordinate systems to locate points on the patterns. Initially I will use simple Cartesian coordinates, which will be later extended to linear but non-orthogonal coordinates. These will enable the symmetry operators to be expressed in matrix form and the effect of coordinate transformations on the operators to be deduced. At the end of this part, the reader should be capable of constructing graphically the 17 planar "wallpaper" groups from their Hermann–Mauguin symbols (Hermann and Mauguin, 1935) and to understand all the entries in the ITC related to them (Volume A, pages 91–111), with the exception of the reflection conditions. A significant amount of space will be subsequently devoted to the 230 space groups and their various settings, but only minimal additions to the concepts and notation already introduced will be needed. The final part of the book will be devoted to a brief overview of symmetry in reciprocal space, with particular reference to the deduction of the reflection conditions.

Acknowledgements

I am deeply grateful to Michael Grazer, Dimitri Argyriou and Laurent Chapon for their thorough reading of the manuscript and for their comments and suggestions.

Figure acknowledgements

Figures 3.16, 7.1 and 7.2: M. C. Escher's *Symmetry Drawing E69* © 2007 The M. C. Escher Company, Holland. All rights reserved. www.mcescher.com.

Figure 3.17: M. C. Escher's *Symmetry Drawing E53* © 2007 The M. C. Escher Company, Holland. All rights reserved. www.mcescher.com.

Contents

PART I SYMMETRY GROUPS: AN INTRODUCTION

1 Symmetry around a fixed point — 3
  1.1 The symmetries of a parallelogram, an arrow and a rectangle — 3
  1.2 What do the graphic symbols for **2** and *m* really mean? Graphs and their symmetry — 4
  1.3 Symmetry operators — 4
  1.4 Group structure: a few formal definitions — 6
  1.5 Multiplication tables for simple groups — 8
  1.6 Graph symmetry vs. composition — 9
  1.7 Conjugation and conjugation classes — 10
  1.8 Rules to obtain 2D multiplication tables — 10
  1.9 The remaining 2D point groups — 11
  1.10 Graph symmetry, conjugation and patterns — 11
  1.11 The 2D point groups in the ITC — 12
  1.12 Group–subgroup relations — 17

2 Frieze patterns and frieze groups — 19
  2.1 Symbols for frieze groups — 22
  2.2 Operator graphs generalized to frieze patterns — 22
  2.3 More formal concepts for frieze groups — 23
  2.4 Relation between graph symmetry and composition — 24
  2.5 Why did we introduce graphs in the first place? Graphical multiplication tables — 24
  2.6 A graphical multiplication table for frieze patterns — 25
  2.7 Commutation: how to "switch" operators — 27
  2.8 Introducing an origin: normal form for operators — 28
  2.9 Choice of origin — 29
  2.10 Composition of operators in normal form — 30
  2.11 Point groups and local symmetry — 30
  2.12 Coordinate systems for frieze groups — 30
  2.13 Frieze groups in the ITC — 31

3 Wallpaper (plane) groups — 35
  3.1 Transformation of translations by graph symmetry — 35
  3.2 The symmetry of the translation set — 36
  3.3 Lattices — 36
  3.4 Crystallographic restriction — 36

	3.5 Bravais lattices in 2D	38
	3.6 Unit cells in 2D	40
	3.7 Composition rules in 2D	41
	3.8 Graphical multiplication table in 2D	43
	3.9 Nomenclature for wallpaper groups	43
	3.10 The 17 wallpaper groups	44
	3.11 Analyzing wallpaper and other 2D art using wallpaper groups	45
	3.12 Constructing wallpaper groups	48

PART II CRYSTALLOGRAPHIC COORDINATE SYSTEMS

4	**Coordinate systems in crystallography**	**53**
	4.1 Change of origin and generalized coordinate transformations	56
	4.2 The most general coordinate transformation	56
5	**The mathematical form of symmetry operators**	**57**
	5.1 Symmetry operators in Cartesian coordinates	57
	5.2 Rotation matrices	58
	5.3 Symmetry operators in generalized crystallographic coordinates	59
6	**Distances, angles and the real and reciprocal spaces**	**61**
	6.1 Determination of distances and angles in real space: the metric tensor	61
	6.2 Dual basis and coordinates: the reciprocal space	62
	6.3 An auxiliary Cartesian system	65
7	**A phase transition in two dimensions**	**67**
	7.1 Low-symmetry group	67
	7.2 Wyckoff positions	67
	7.3 Basis transformation	69
	7.4 Metric tensor	70

PART III SYMMETRY IN THREE DIMENSIONS

8	**Point groups in 3D**	**75**
	8.1 Generalized rotations in 3D	75
	8.2 3D point groups with a 2D point-group projection	76
	8.3 Other 3D point groups: the five cubic groups	79
	8.4 3D point groups in the ITC	80
9	**The 14 3D Bravais lattices**	**83**
	9.1 Introduction	83
	9.2 Construction of the 14 Bravais lattices	83
10	**3D space-group symmetry**	**89**
	10.1 Roto-translations in 3D	89

10.2	Glide planes in 3D	90
10.3	Graphical notation for 3D symmetry operators	91
10.4	Multiplication tables in 3D	92
10.5	Construction of space groups in 3D	93

PART IV RECIPROCAL SPACE

11 Symmetry and reflection conditions in reciprocal space — 101
11.1 Real and reciprocal lattice: recapitulation — 101
11.2 Reciprocal lattice – an alternative definition — 102
11.3 Centering extinctions — 103
11.4 Holohedry of the reciprocal lattice — 105
11.5 Fourier transform of lattice functions: the "weighed" Reciprocal Lattice and its symmetry — 107
11.6 Further exploitation of the RL symmetry — 113

12 The Wigner–Seitz constructions and the Brillouin zones — 117
12.1 The Wigner–Seitz construction at the origin of reciprocal space — 118
12.2 The extended Wigner–Seitz scheme: construction of higher-order Brillouin zones — 120

References — 123
Index — 125

SYMMETRY GROUPS: AN INTRODUCTION

Part I

Concepts of symmetry are of capital importance in all branches of the physical sciences. In physics, *continuous* symmetry is particularly important and is rightly emphasized because of its connection with conserved quantities through the famous Noether's theorem, which, in extreme synthesis, states that for every continuous symmetry there exists a corresponding conserved quantity. For example, translational invariance of the Hamiltonian implies the conservation of linear momentum, rotational invariance that of angular momentum, and so on.

Discrete symmetries – those in which a figure of a solid is invariant on rotation by a finite angle, on reflection and/or on translation by a finite vector – are also very familiar to us. Symmetry is found everywhere in nature, particularly in connection with the crystalline state. Far from being simply a descriptive tool for crystal structures, discrete symmetries drive some of the most profound insights about the properties of crystals, for example, through Neumann's principle and the Curie asymmetry principle. In the first part of this book, we will gradually familiarize ourselves with discrete symmetry transformations. We will start with *finite* groups, such as those describing the discrete symmetries around a fixed point. The so-called "frieze" group will provide the simplest examples of infinite discrete symmetries in one dimension. Finally, we will describe the symmetries of two-dimensional (2D) periodic patterns, the so-called "wallpaper" groups. These display all the important features of three-dimensional (3D) space groups, and have the advantage of being manageable and simple to draw, so that a motivated student can expect to be able to construct them from first principles.

Symmetry around a fixed point

1

In this chapter, we will introduce a few basic symmetry concepts by describing a few simple transformations of a 2D pattern around a fixed point. The transformations we are interested in are *discrete* (i.e., we are not interested in infinitesimal transformations) and preserve distances (isometric transformations). In essence, the transformations in question are *rotations* around the fixed point by a rational fraction of 360°, reflections by a line (by analogy with 3D, we will often call this a "plane") passing through the fixed point and combinations thereof. As we shall see later on, the very concept of "transformation" (or "operation", an equivalent term that we will introduce shortly) will require some clarification. To begin with, a few simple and intuitive examples should serve to introduce the basic concepts employed in this chapter.

1.1 The symmetries of a parallelogram, an arrow and a rectangle

A parallelogram has two-fold (180°) rotational symmetry around its center. We will denote the two-fold axis with a vertical "pointy" ellipse (Figure 1.1, left) and with the number 2. An arrow is symmetric on reflection through a line through its middle. We will denote this reflection with a thick line (Figure 1.1, right) and with the letter *m*.

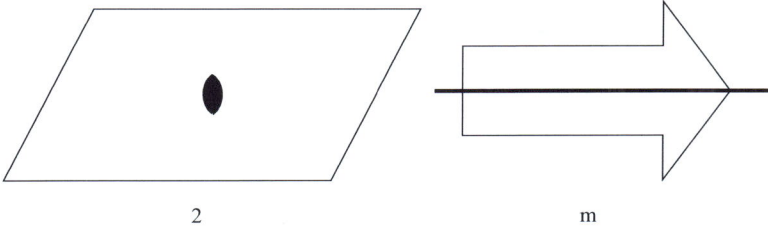

Fig. 1.1 The symmetry of a parallelogram (left) and of an arrow (right).

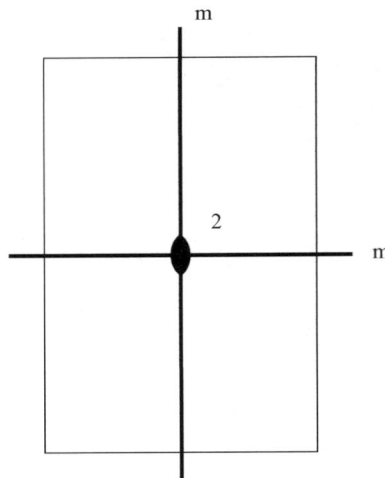

Fig. 1.2 The symmetry of a rectangle.

In the previous cases, the transformations 2 or m are the only ones present, if one excludes the trivial identity transformation. However, these two transformations can also be found combined in the case of the *rectangle*. Here, we have two m transformations and a 2 transformation at their intersection, which is also the fixed point of the figure (Figure 1.2).

1.2 What do the graphic symbols for 2 and m really mean? Graphs and their symmetry

By inspecting Figure 1.2, it is easy to understand that the graphic symbols (or "graphs") that we have drawn represent *sets of invariant points* – this definition seems to have been first introduced by Haüy (1822). The points on the m graphs are transformed into themselves by the reflection, whereas all the other points are transformed into different points. The same is true for 2, for which the graph is also the "fixed point" that, in this type of symmetry, is left invariant by all transformations. Since graphs are part of the pattern, it is natural that they should also be subject to transformations, unless they coincide with the fixed point.[1] This is clear by looking at Figure 1.3, which represents the symmetry of a square. The central symbol, on the fixed point represents three transformations: the counterclockwise rotation by 90° (4^+), the clockwise rotation by 90° (4^-) and the rotation by 180° (2). All the other transformations are of type m. It is easy to see that for each mirror line m there is another line rotated by 90°, which can be thought of as its symmetry partner via the transformations 4^+ or 4^-. The two 90°-rotations are (perhaps less obviously) the mirror image of each other. However, it is also clear that the planes rotated by 45° cannot be obtained by symmetry from the other elements. Where do they come from? We anticipate the answer here: the diagonal mirrors are obtained by successive application (we will later call this *composition* or *multiplication*) of a horizontal (or vertical) mirror followed by a 90° rotation. In order to understand this, we need to learn a little more about these transformations and their relations.

[1] Throughout this book, the words "graph of an operator" and "symmetry element" are used to describe sets of invariant points under a certain transformation. The word "element" also indicates a member of a set – in particular a "symmetry operator" is an "element" of a *symmetry group*. The reader should be careful to avoid confusion, although the meaning is usually clear from the context.

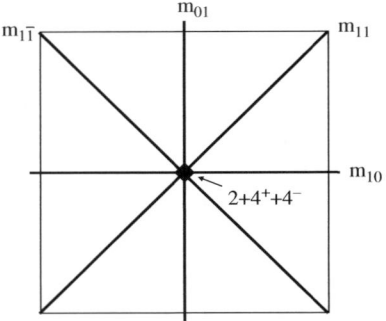

Fig. 1.3 The symmetry of a square. The central symbol describes three transformations: "4^+" (counterclockwise rotation by 90°), "4^-" (clockwise rotation by 90°) and "2" (rotation by 180°). The labeling of the four mirror lines is referred to in the text.

1.3 Symmetry operators

In a somewhat more formal way, the transformations we described as 2, m, 4^+, etc., are said to be produced by the application of *symmetry operators*. These transformations can be interpreted in more than one way, and this is the source of much confusion; it is therefore worth discussing them in some detail. Operators of this kind define a two-way (one-to-one) correspondence of the plane (or space) into itself. In other words, each point p is uniquely associated by the transformation with a new point p'. Likewise, each point q', after the transformation, will receive the attributes of a point q. As already mentioned, here we are only concerned with *isometries*, i.e., operators that preserve distances (and therefore shapes). In other words, the distance between points p' and q' is the same as for points p and q, and likewise for angles. In the interpretation employed in this book, the *identity* of the points themselves (and their coordinates – see later on) are unchanged by the transformation, but the *attributes* of point p are transferred to p'. This is known as an *active transformation* (in an alternative interpretation, a *passive transformation* transforms the axis and consequently the coordinates). Examples of *attributes* in this context are the color or the relief of the pattern, etc. In crystallography, content will mean an atom, a magnetic moment, a vector or tensor quantity, etc. So, in summary:

- **Operators**: Transform (move) the whole pattern (i.e., the *attributes*, or content, of all points in space). We denote operators in *italic font* and we used parentheses ( ) around them for clarity, if required.
- **Symmetry operators**: A generic operator as described above is said to be a *symmetry operator* if, upon transformation, the new pattern is indistinguishable from the original one. Let us imagine that a given operator *g* transforms point p to point p'. In order for *g* to be a *symmetry* operator, the *attributes* of the two points must be in some sense "the same". This is illustrated in a general way in Figure 1.4. **Note:** the inquisitive reader may at this point appreciate the difference between *scalar* attributes, such as color or relief, which, taken in isolation, are left unchanged by the transformation, and more complex attributes (e.g., vectors) which can themselves be

Fig. 1.4 A generic symmetry operator acting on a pattern fragment.

rotated or mirrored. Even in these more complex examples, the same principles apply: the operator sets up a point-by-point correspondence and, in the case of a symmetry operator, the pattern is unchanged after the attributes are transformed as appropriate and transferred to the new points.

- **Application of operators** to points or parts of the pattern, relating them to other points or sets of points. We indicate this with the notation v = gu, where u and v are sets of points. We denote sets of points with Roman fonts and put square brackets [ ] around them for clarity, if required. We will also say that pattern fragment u is *transformed* by g into pattern fragment v. If the pattern is to be symmetric, v must have the same attributes as u in the sense explained above.

- **Operator composition**: This is the sequential ordered application of two operators, and we indicate this with $g \circ h$. The new operator thus generated acts as $(g \circ h)$u = $g[h$u$]$. With our conventions, *the operator to the **right** (h in this case) is applied **first***. **Important note**: Symmetry operators in general *do not commute*, so the order is important. We will see later on that translations (which are represented by vectors) can be symmetry operators. The composition of two translations is simply their *vector sum*.

- **Operator graphs**: These are *sets of points* in space that are invariant (i.e., are transformed into themselves) upon the application of a given operator. We draw graphs with conventional symbols indicating how the operator acts. We denote the graph of the operator g (i.e., the invariant points) as [g]. **Note**: graphs can be thought of as parts of the pattern, and are subject to symmetry like everything else (as explained above). Sometimes, as in the above case of a four-fold axis, the graphs of two distinct operators coincide (e.g., left and right rotations around the same axis). In this case, the conventional symbol will account for this fact.

1.4 Group structure: a few formal definitions

1.4.1 The mathematical structure of a group

Sets of symmetry operators of interest for crystallography have the mathematical structure of a *group*. The first use of the term "group" in this context, as well as the development of a connections between "groups" and "field theory" are due to the French mathematician Évariste Gaulois (1811–1832). The group structure is defined by:

- A binary operation (usually called **composition** or **multiplication**) must be defined. We indicate this with the symbol "$\circ$".
- Composition must be **associative**: for every three elements f, g and h of the set,

$$f \circ (g \circ h) = (f \circ g) \circ h. \qquad (1.1)$$

- A "neutral element" (i.e., the identity, usually indicated with E) must exist, so that for every element g:

$$g \circ E = E \circ g = g. \qquad (1.2)$$

- Each element g has an **inverse** element g^{-1} so that

$$g \circ g^{-1} = g^{-1} \circ g = E. \qquad (1.3)$$

Problem 1.1: using the associative property, show that the right and left inverses always coincide.

1.4.2 Some more nomenclature

- **Order**: two meanings. The *order of a group* is the number of elements in the group. All the groups we have seen so far have a finite number of elements. By contrast, the *order of an element* is the number of times the operator element must be composed with itself (the power) to obtain the identity. Of the symmetry operators we have seen so far, the mirror and the two-fold axis have order 2, whereas the four-fold axes have order 4.
- **Generators**. A set of operators (not necessarily minimal or unique) that generates the whole group by composition. All groups of interest for us have a finite number of generators.

Problem 1.2: determine sets of generators for the groups in Figures 1.1–1.3. Some have a single generator.

- **Points of special symmetry**. These are points that are left invariant by the application of one or more operators, so they have fewer "equivalent points" than a general point. In the case of point groups, they coincide with the graphs of the symmetry operators.
- **Multiplication table**. A square table (order × order) defining the composition rules of a group (i.e., every element composed with every other to the right and to the left). Two groups having the same multiplication table are said to be the same *abstract group*.
- **Subgroups**. A subset of a group that is also a group. Clearly, the subgroup at the very least has to contain the identity and the inverse of each of its elements.
- **Point groups**. The groups describing discrete symmetries around a fixed point are known as *point groups*. Clearly, there are an infinite number of point groups (think, for example, of the symmetry groups of each regular polygon). However, as we shall see later, there are only 10 *crystallographic point groups* in 2D and 32 in 3D.

1.5 Multiplication tables for simple groups

If a finite group G has n elements, then clearly there will be n^2 possible multiplications in the group. These can be collected in the form of an $n \times n$ matrix, known as the **multiplication table**.

The **parallelogram** and **arrow groups** (Figure 1.1) have the same multiplication table, shown in Table 1.1, so they are the same *abstract group*. They have only two elements: 2 or m and the identity, which we will indicate with E. The **rectangle group** (Figure 1.2) has four elements, and its multiplication table is shown in Table 1.2.

The **square group** (Figure 1.3) has eight elements, and its multiplication table is shown in Table 1.3. Note that here the *order* of the operators is important. We will apply *first* the operators on the top, *then* those on

Table 1.1 Multiplication table for the symmetry groups of the parallelogram and of the arrow (Figure 1.1). There are only two elements, the identity E and the two-fold rotation 2 or the mirror line m.

	E	2 or m
E	E	2 or m
2 or m	2 or m	E

Table 1.2 Multiplication table for the symmetry group of the rectangle. There are four elements, the identity E, two orthogonal mirror planes m_{10} and m_{01} and the two-fold rotation 2.

	E	m_{10}	m_{01}	2
E	E	m_{10}	m_{01}	2
m_{10}	m_{10}	E	2	m_{01}
m_{01}	m_{01}	2	E	m_{10}
2	2	m_{01}	m_{10}	E

Table 1.3 Multiplication table for the symmetry group of the square. There are eight elements, the identity E, four mirror planes orthogonal in pairs m_{10}, m_{01}, m_{11} and $m_{1\bar{1}}$, the two-fold rotation 2 and two rotations by 90° in the positive (4^+) and negative (4^-) directions.

	E	m_{10}	m_{01}	m_{11}	$m_{1\bar{1}}$	2	4^+	4^-
E	E	m_{10}	m_{01}	m_{11}	$m_{1\bar{1}}$	2	4^+	4^-
m_{10}	m_{10}	E	2	4^-	4^+	m_{01}	$m_{1\bar{1}}$	m_{11}
m_{01}	m_{01}	2	E	4^+	4^-	m_{10}	m_{11}	$m_{1\bar{1}}$
m_{11}	m_{11}	4^+	4^-	E	2	$m_{1\bar{1}}$	m_{10}	m_{01}
$m_{1\bar{1}}$	$m_{1\bar{1}}$	4^-	4^+	2	E	m_{11}	m_{01}	m_{10}
2	2	m_{01}	m_{10}	$m_{1\bar{1}}$	m_{11}	E	4^-	4^+
4^+	4^+	m_{11}	$m_{1\bar{1}}$	m_{01}	m_{10}	4^-	2	E
4^-	4^-	$m_{1\bar{1}}$	m_{11}	m_{10}	m_{01}	4^+	E	2

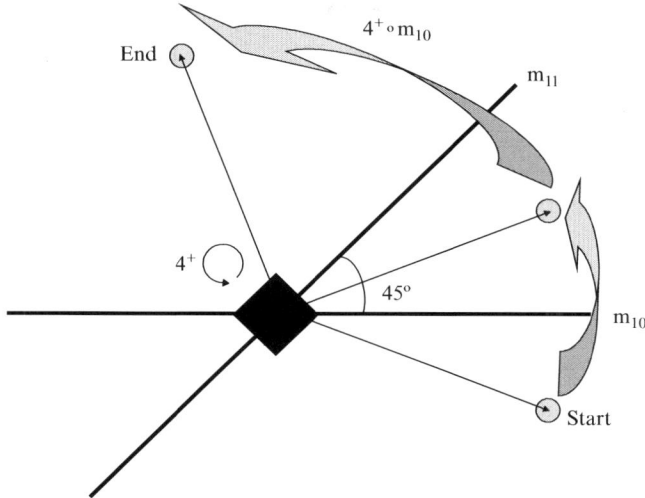

Fig. 1.5 A graphical illustration of the composition of the operators 4^+ and m_{10} to give $4^+ \circ m_{10} = m_{11}$. Note that the operator to the *right* (in this case m_{10}) is applied *first*.

the side. It is easy to see that some of the elements do not commute – for instance the four-fold axes with the mirror planes.

Figure 1.5 illustrates in a graphical way the composition of the operators 4^+ and m. The fragment to be transformed is indicated with "start", and the two operators are applied in order one after the other, until one reaches the "end" position. It is clear by inspection that "start" and "end" are related by the "diagonal mirror" operator m_{11}, as we anticipated.

Note that the two operators 4^+ and m_{10} *do not commute* (see Table 1.3):

$$4^+ \circ m_{10} = m_{11}$$
$$m_{10} \circ 4^+ = m_{1\bar{1}}. \quad (1.4)$$

1.6 Graph symmetry vs. composition

Let us now return to the issue of why some apparently equivalent symmetry elements, such as the two mirror lines in Figure 1.2, are not related by symmetry. By looking at Table 1.2, we see that the orthogonal planes are obtained by *composition with the two-fold axis*. We see here, for the first time, that composing two operators is not the same thing as applying one operator to the graph of the other. Likewise, by looking at Table 1.3, we see that two orthogonal planes (say, m_{10} and m_{01}), which are clearly related by a 90° rotation, are not obtained by simple composition with the 4^+ or 4^- operators. However, we also know that we must be able to generate the new operators from the old ones by some form of composition. So what is the composition corresponding to a given graph symmetry operation? The answer is the subject of the next problem:

Problem 1.3: prove that the following equation is valid and verify it with Table 1.2:

$$g[h] = [g \circ h \circ g^{-1}]. \tag{1.5}$$

We read Eqn. (1.5) in the following way: "The graph of the operator h transformed with the operator g is equal to the graph of the operator $g \circ h \circ g^{-1}$". This relation clearly shows that graph symmetry is *not* equivalent to composition.

1.7 Conjugation and conjugation classes

The group operation we just introduced, $g \circ h \circ g^{-1}$, also has a special name – it is known as **conjugation**. If $k = g \circ h \circ g^{-1}$ we say that "k and h are *conjugated* through the operator g".

Problem 1.4: prove that conjugation is reflexive (i.e., g is conjugated with itself), symmetric (i.e., if g is conjugated with f, f is conjugated with g) and transitive (i.e., if g is conjugated with f and f is conjugated with h then g is conjugated with h), and is therefore an equivalence relation.

Since conjugation is an equivalence relation, it defines disjoint classes of operators related to each other (and to no other operator) by conjugation. These are known as **conjugation classes**. In a less formal way, and remembering how we arrived at the definition of conjugation, we can say that **conjugation classes group together operators with symmetry-related graphs**.

Problem 1.5: analyze the square group (Figure 1.3 and Table 1.3) and show that it has five conjugation classes – two with a single element and three with two elements each. In the process, show that the identity always forms a conjugation class with a single element.

Problem 1.6: show that each element of a commutative (Abelian) group is in a conjugation class of its own, i.e., there are as many conjugation classes as elements of the group

1.8 Rules to obtain 2D multiplication tables

It is already clear at this point that multiplication tables can be rather complex to handle, even when the group has only eight elements. The largest 3D crystallographic point group has 48 elements, so its multiplication table has 2304 elements, clearly not a very practical tool.

However, *all* the 2D point-group multiplication tables, including the ones we have not yet seen, can be obtained from three simple rules:

Rule 1 The composition of two rotations (around the same axis) is a rotation by the sum of the angles. Rotations around the vertical axis commute.

Rule 2 The composition of two intersecting planes is a rotation around the intersection. The rotation *angle* is *twice* the angle between the planes. The *direction* of the rotation is from the plane that is applied first (i.e., that appears to the *right* in the composition). From this, it follows that two mirror planes *anticommute*.

Rule 3 This is the reverse of Rule 2. The composition of a plane with a rotation about an axis in the plane itself (in the order $n^+ \circ m$) is a plane obtained by rotating the first plane around the axis by *half* the rotation angle. If the two operators are exchanged, the rotation is in the opposite direction. Note that this is a generalization of what is shown in Figure 1.5.

1.9 The remaining 2D point groups

We have so far encountered *four* 2D point groups. A fifth is the trivial group in which the only symmetry is the identity E. There are five more *crystallographic* 2D point groups, that can be easily obtained using the rules listed above. There are only two new operators, in addition to the ones we know already: the three-fold axis (▲) and the six-fold axis (●). We shall see later why five-fold axes and axes of higher order are not allowed in crystallography. The *ten* 2D point groups can be obtained by using each of the six allowed operators as a *single* generator, or by using each of the four axes and the mirror line m as the two generators.

Problem 1.7: construct the following 2D point groups: (1) four-fold axis without mirror lines (see Figure 1.6); (2) three-fold axis with and without mirror lines (see Figure 1.7 for the latter); (3) six-fold axis with and without mirror lines.

1.10 Graph symmetry, conjugation and patterns

It is often quite easy to spot the parts of the patterns that lie on symmetry graphs. For example, the "spikes" on the snowflake shown in Figure 1.8 correspond to mirror planes in its symmetry. We should note, however, that there are two types of spikes, each occurring six times. This is because, as we recall, there are two types of mirror planes

Fig. 1.6 The central square of this Roman mosaic from Antioch has four-fold symmetry without mirror lines. From (Hayes, 2011).

Fig. 1.7 Two examples of three-fold symmetry without mirror lines. Left: a slightly modified "Recycle" logo. Right: the "Triskele" appears in the symbol of Sicily, but variations of it are a very common element in Celtic art.

(two *conjugation classes*, which are *not* related by symmetry, marked "1" and "2" on the drawing). This example underlines the importance of conjugation classes – as we said before, the graphs of operators in the same conjugation class always carry the same attributes (patterns).

1.11 The 2D point groups in the ITC

1.11.1 Symmetry directions: the key to understanding the ITC

The group symbols used in the ITC employ the so-called **Hermann–Mauguin notation**. The symbols are constructed with letters and numbers in a particular sequence – for example, *6mm* is a *point-group* symbol and $I4_1/amd$ is an ITC *space-group* symbol. This notation is complete and completely unambiguous, and should enable one, with some practice, to construct all the symmetry operator graphs. Nevertheless, the ITC symbols are the source of much confusion for beginners (and even some practitioners). In the following paragraphs we will explain in some detail the point-group notation of the ITC, but here it is perhaps useful to make some general remarks just by looking at the snowflake and its symmetry group diagram (6*mm*) in Figure 1.8.

- The principal symmetry feature of the 6*mm* symmetry is the six-fold axis. Axes with order higher than 2 (i.e., 3, 4 and 6) **define the *primary symmetry direction* and always come up front in the point-group symbol, and right after the lattice symbol (P, I, F, etc.) in the space-group symbols**. This is the meaning of the first character in the symbol 6*mm*.
- The next important features are the **mirror planes**. We can pick any plane we want and use it to define the *secondary symmetry direction*. For example, in Figure 1.8, we could define the secondary symmetry direction to be horizontal and perpendicular to the vertical mirror plane marked "1". This is the meaning of the second character in the symbol 6*mm*.
- The *tertiary symmetry direction* **is never symmetry equivalent to the other two**. In other words, the operator "*m*" appearing in the

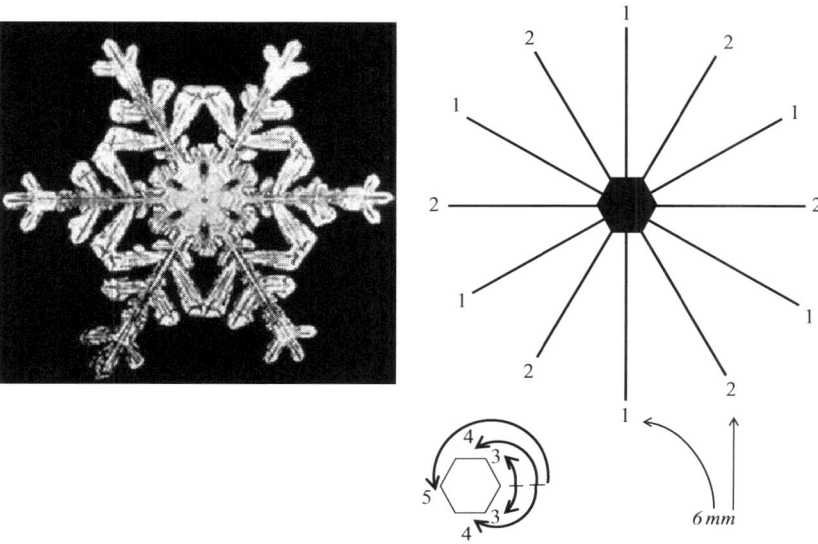

Fig. 1.8 Left: A snowflake by Vermont scientist-artist Wilson Bentley, c. 1902. Right: The symmetry group of the snowflake, 6*mm* in the ITC notation. The group has six classes, five marked on the drawing plus the identity operator E. Note that there are two classes of mirror planes, marked "1" and "2" on the drawing. One can see on the snowflake picture that their graphs contain *different patterns*.

third position as **6*mm*** **does not belong to the same class as either of the other two symbols**. It has therefore necessarily to refer to a mirror plane of the *other class*, marked as "2".

- Therefore, **in 6*mm*, secondary and tertiary symmetry directions make an angle of** $30°$ **with each other**. Likewise **in 4*mm*, (the square group) secondary and tertiary symmetry directions make an angle of** $45°$ **with each other**.

Operators listed in the ITC group symbols *never belong to the same conjugation class.*

1.11.2 Detailed description of the 2D point-group tables in the ITC

The *ten* 2D point groups are listed in ITC, Volume A (Hahn, 2002) on pages 768–769 (Table 10.1.2.1 therein, see Figure 1.9). We have not introduced all the notation at this point, but it is worth examining the entries in some detail, as the principles of the notation will be largely the same throughout the ITC.

- **Reference frame**: All point groups are represented on a circle with thin lines through it. The fixed point is at the center of the circle. All symmetry-related points are at the same distance from the center (remember that symmetry operators are isometries, i.e., they preserve distances), so the circle around the center locates symmetry-related points. The thin lines represent possible systems of coordinate axes (*crystal axes*) to locate the points. We have not introduced axes at this point, but we will note that the lines have the same symmetry of the pattern.
- **System**: Once again, this refers to the type of axes and choice of unit length. The classification is straightforward.
- **Point-group symbol**: This is listed in the top left corner, and generally consists of three characters: a number followed by two letters (such as 6*mm*). When there is no symmetry along a particular direction (see below), the symbol is omitted, but it could also be replaced by a "1". For example, the point group *m* can be also written as 1*m*1. The first symbol stands for one of the allowed rotation axes perpendicular to the sheet. Each of the other two symbols represents symmetry elements defined by *inequivalent* symmetry directions, known as "secondary" and "tertiary", respectively. In this case, they are *sets of mirror lines that are equivalent by rotational symmetry* or, in short, different conjugation classes. The lines associated with each symbol are not symmetry-equivalent (so they belong to different conjugation classes). For example, in the point group 4*mm*, the first *m* stands for two orthogonal mirror lines. The second *m* stands for two other (symmetry-inequivalent) orthogonal mirror lines rotated by $45°$ with respect to the first set. Note that all the symmetry directions are equivalent for the three-fold axis 3, so

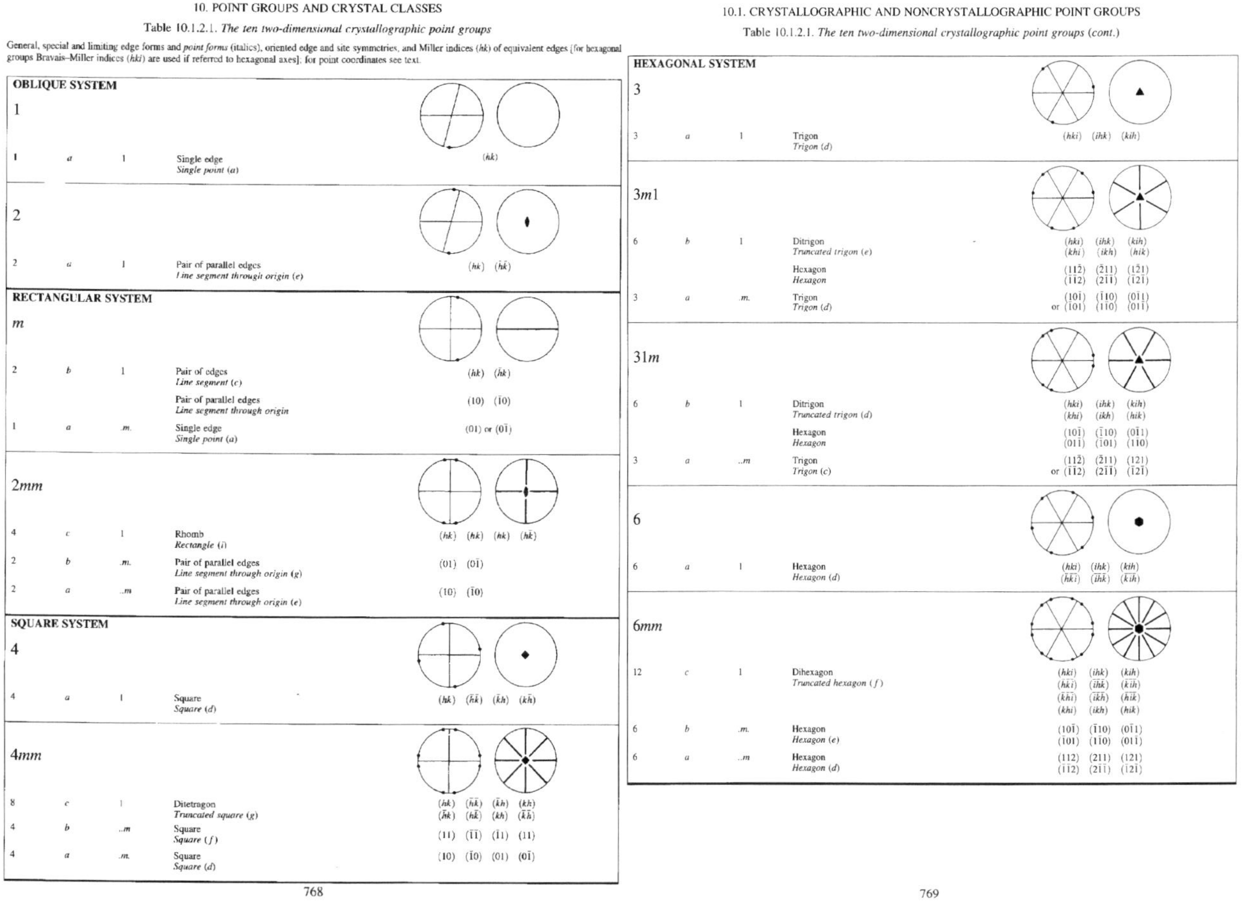

10. POINT GROUPS AND CRYSTAL CLASSES

Table 10.1.2.1. *The ten two-dimensional crystallographic point groups*

General, special and limiting edge forms and *point forms* (italics), oriented edge and site symmetries, and Miller indices (*hk*) of equivalent edges [for hexagonal groups Bravais–Miller indices (*hki*) are used if referred to hexagonal axes]; for point coordinates see text.

OBLIQUE SYSTEM				
1				
1	*a*	1	Single edge *Single point (a)*	(hk)
2				
2	*a*	1	Pair of parallel edges *Line segment through origin (e)*	(hk) $(\bar{h}\bar{k})$
RECTANGULAR SYSTEM				
m				
2	*b*	1	Pair of edges *Line segment (c)*	(hk) $(\bar{h}k)$
			Pair of parallel edges *Line segment through origin*	(10) $(\bar{1}0)$
1	*a*	.*m*.	Single edge *Single point (a)*	(01) or $(0\bar{1})$
2mm				
4	*c*	1	Rhomb *Rectangle (i)*	(hk) $(\bar{h}\bar{k})$ $(\bar{h}k)$ $(h\bar{k})$
2	*b*	.*m*.	Pair of parallel edges *Line segment through origin (g)*	(01) $(0\bar{1})$
2	*a*	..*m*	Pair of parallel edges *Line segment through origin (e)*	(10) $(\bar{1}0)$
SQUARE SYSTEM				
4				
4	*a*	1	Square *Square (d)*	(hk) $(\bar{h}\bar{k})$ $(\bar{k}h)$ $(k\bar{h})$
4mm				
8	*c*	1	Ditetragon *Truncated square (g)*	(hk) $(\bar{h}\bar{k})$ $(\bar{k}h)$ $(k\bar{h})$ $(\bar{h}k)$ $(h\bar{k})$ (kh) $(\bar{k}\bar{h})$
4	*b*	..*m*	Square *Square (f)*	(11) $(\bar{1}\bar{1})$ $(\bar{1}1)$ $(1\bar{1})$
4	*a*	.*m*.	Square *Square (d)*	(10) $(\bar{1}0)$ (01) $(0\bar{1})$

768

10.1. CRYSTALLOGRAPHIC AND NONCRYSTALLOGRAPHIC POINT GROUPS

Table 10.1.2.1. *The ten two-dimensional crystallographic point groups (cont.)*

HEXAGONAL SYSTEM				
3				
3	*a*	1	Trigon *Trigon (d)*	(hki) (ihk) (kih)
3m1				
6	*b*	1	Ditrigon *Truncated trigon (e)*	(hki) (ihk) (kih) (khi) (ikh) (hik)
			Hexagon *Hexagon*	$(11\bar{2})$ $(\bar{2}11)$ $(1\bar{2}1)$ $(\bar{1}\bar{1}2)$ $(2\bar{1}\bar{1})$ $(\bar{1}2\bar{1})$
3	*a*	.*m*.	Trigon *Trigon (d)*	$(10\bar{1})$ $(\bar{1}10)$ $(0\bar{1}1)$ or $(\bar{1}01)$ $(1\bar{1}0)$ $(01\bar{1})$
31m				
6	*b*	1	Ditrigon *Truncated trigon (d)*	(hki) (ihk) (kih) (khi) (ikh) (hik)
			Hexagon *Hexagon*	$(10\bar{1})$ $(\bar{1}10)$ $(0\bar{1}1)$ $(01\bar{1})$ $(\bar{1}01)$ $(1\bar{1}0)$
3	*a*	..*m*	Trigon *Trigon (c)*	$(11\bar{2})$ $(\bar{2}11)$ (121) or $(\bar{1}\bar{1}2)$ $(2\bar{1}\bar{1})$ $(\bar{1}2\bar{1})$
6				
6	*a*	1	Hexagon *Hexagon (d)*	(hki) (ihk) (kih) $(\bar{h}\bar{k}\bar{i})$ $(\bar{i}\bar{h}\bar{k})$ $(\bar{k}\bar{i}\bar{h})$
6mm				
12	*c*	1	Dihexagon *Truncated hexagon (f)*	(hki) (ihk) (kih) $(\bar{h}\bar{k}\bar{i})$ $(\bar{i}\bar{h}\bar{k})$ $(\bar{k}\bar{i}\bar{h})$ $(\bar{k}\bar{h}\bar{i})$ $(\bar{i}\bar{k}\bar{h})$ $(\bar{h}\bar{i}\bar{k})$ (khi) (ikh) (hik)
6	*b*	.*m*.	Hexagon *Hexagon (e)*	$(10\bar{1})$ $(\bar{1}10)$ $(0\bar{1}1)$ $(\bar{1}01)$ $(1\bar{1}0)$ $(01\bar{1})$
6	*a*	..*m*	Hexagon *Hexagon (d)*	(112) (211) (121) $(\bar{1}\bar{1}2)$ $(2\bar{1}\bar{1})$ $(\bar{1}2\bar{1})$

769

Fig. 1.9 Two-dimensional point groups: a reproduction of pages 768–769 of the ITC (Hahn, 2002).

either the secondary or the tertiary direction must carry a "1" (see below).
- **General and special positions**: Below the point-group symbol, we find a list of general and special positions (points), the latter lying on a symmetry element, and therefore having fewer "equivalent points". Note that the unique point at the center is always omitted. From left to right, we find:

  Column 1 The *multiplicity*, i.e., the number of equivalent points.

  Column 2 The Wyckoff letter (Wyckoff, 1922) (historical references are from Lima-de Faria, 1990), starting with *a* from the bottom up. Symmetry-inequivalent points with the same symmetry (i.e., lying on symmetry elements of the same type) are assigned different letters.

  Column 3 The site symmetry, i.e., the symmetry element (always a mirror line for 2D) on which the point lies. Dots are used to indicate which symmetry element in the point-group symbol one refers to. For example, site *b* of point group 4*mm* has symmetry ..*m*, i.e., lies on the *second* set of mirror lines, at 45° from the first set.

  Column 4 Name of crystal and *point* forms (the latter in italic) and their "limiting" (or degenerate) forms. Point forms are easily understood as the polygon (or later polyhedron) defined by sets of equivalent points with a given site symmetry. Crystal forms are historically more important, because they are related to *crystal shapes*. They represent the polygon (or polyhedron) with sides (or faces) passing through a given point of symmetry and orthogonal to the radius of the circle (sphere). We shall not be further concerned with forms.

  Column 5 Miller indices. For point groups, Miller indices are best understood as related to crystal forms, and represent the inverse intercepts along the crystal axes. By the well-known "law of rational indices", real crystal faces are represented by integral Miller indices. We also note that for the hexagonal system three Miller indices (and three crystal axes) are shown, although naturally only two are needed to define coordinates. The "Miller" indices (Miller, 1839) were actually introduced 15 years earlier by Whewell (1825).

- **Projections**: For each point group, two diagrams are shown. It is worth noting that for 3D point groups, these diagrams are *stereographic projections* of systems of equivalent points. The diagram on the **left** shows the projection circle, the crystal axes as thin lines, and a set of equivalent general positions, shown as dots. The diagram on the **right** shows the symmetry elements, using the same notation we have already introduced.

- **Settings**: We note that one of the ten 2D point groups is shown twice with a different notation, 3*m*1 and 31*m*. By inspecting the diagram, it is clear that the two only differ for the position of the

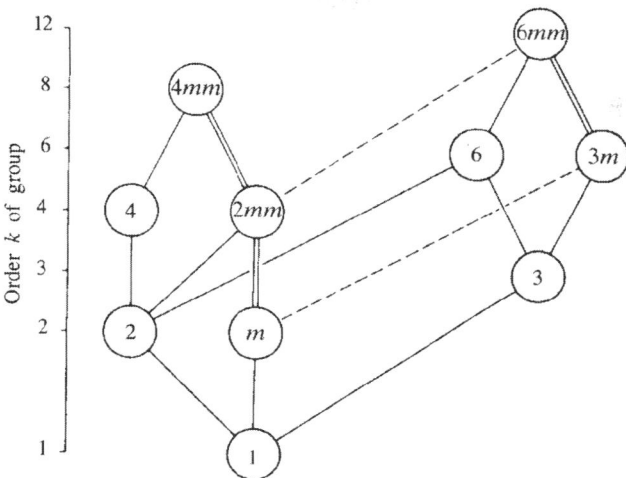

Fig. 1.10 Group–subgroup relations for 2D point groups: a reproduction from page 795 of the ITC (Hahn, 2002).

crystal axes with respect of the symmetry elements. In other words, the difference is entirely conventional, and refers to the choice of axes. We refer to this situation, which recurs throughout the ITC, as two different *settings* of the same point group.

1.12 Group–subgroup relations

The group–subgroup relations for 2D point groups are shown in a diagrammatic form on page 795 in ITC, Volume A, (Hahn, 2002), Figure 10.1.3.1 therein, reproduced in Figure 1.10. The relations are shown in the form of a *family tree*. The *order* of the group (i.e., the number of elements) is shown as a scale on the left side. Lines are shown to connect point groups that differ by a minimal number of operators (known as *maximal subgroups/minimal supergroups*). A single continuous line is shown when a point group has only one subgroup of a given type. Multiple lines are shown when more than one subgroup of a given type exists, but the subgroups are not equivalent by symmetry. A dashed line is shown when the subgroups are equivalent by symmetry. This difference should be clear by inspecting the diagrams of 4*mm* and 6*mm*, both having 2*mm* as a maximal subgroup.

Problem 1.8: by analyzing the graphical representations of 4mm and 6mm, show how many times 2mm is contained in each of them. In particular, show that the two mirror lines in 2mm belong to the same conjugation class in 4mm and to different conjugation classes in 6mm.

Frieze patterns and frieze groups

2

Friezes are two-dimensional patterns that are repetitive in one dimension. They have been employed by essentially all human cultures to create ornamentations on buildings, textiles, metalwork, ceramics, etc. (see examples below). Depending on the nature of the object, these decorative motifs can be linear, circular (as on the neck of a vase) or follow the contour of a polygon. Here, we will imagine that the pattern is unwrapped to a linear strip and is infinite. In addition, we will only consider monochrome patterns. Although the design can comprise a variety of naturalistic or geometrical elements, as far as the symmetry is concerned frieze patterns follow a very simple classification. There are only five types of symmetries, three of them already known to us:

1. **Rotations** about an axis perpendicular to the viewing plane. Only the two-fold rotation, as for the symmetry of the letter "S", is allowed.
2. **Reflections** across lines in the plane of the pattern, **perpendicular to the translations**, as for the symmetry of the letter "V". Again, we will liberally use the term "mirror plane" instead of the more rigorous "mirror line", to be consistent later on with the space-group definitions.
3. **Reflections** across a line in the plane of the pattern, **parallel to the translations**, as for the symmetry of the letter "K".
4. Translations. This is a new symmetry that we did not encounter for point groups, since by definition they had a fixed point, whereas translations leave no point fixed. In all frieze patterns, there exists a fundamental ("primitive") translation that defines the repeated pattern. Its opposite (say, left instead of right) is also a symmetry element, as are all multiples thereof, clearly an infinite number of symmetry translations. As we can see, one very important property of symmetry translations is that they form not only a group, but also a **subset of a linear space**. In other words, **all symmetry translations can be generated as linear combinations**

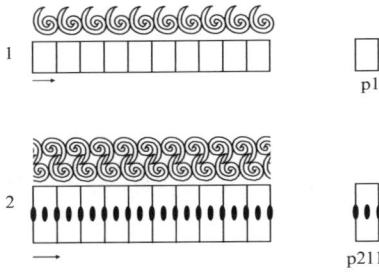

Fig. 2.1 Frieze groups $p1$ and $p211$.

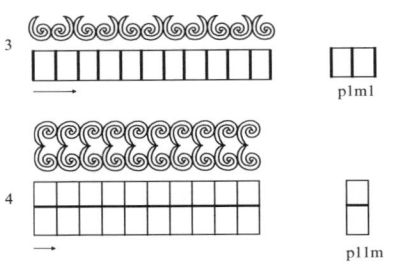

Fig. 2.2 Frieze groups $p1m1$ and $p11m$.

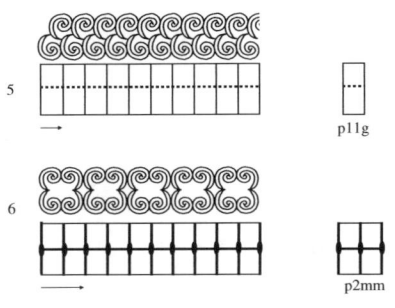

Fig. 2.3 Frieze groups $p11g$ and $p2mm$.

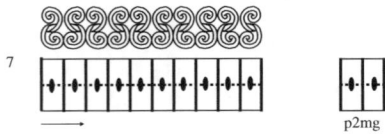

Fig. 2.4 Frieze group $p2mg$.

with integral coefficients of the "primitive" translations, which can be taken as the basis vectors of the linear space. This is a general result, valid in all dimensions.

5. **Glides.** This is a composite symmetry, which combines a translation with a *parallel* reflection, neither of which on its own is a symmetry. The primitive translation is always *twice* the glide translation, for reasons that should be immediately clear (see Problem 2.1 below). This symmetry is represented by the repeated fragment ⌊Γ, as in …⌊Γ⌊Γ⌊Γ….

Problem 2.1: by exploiting the fact that a translation commutes with a mirror line parallel to it and that the square of a mirror line is the identity, show that the glide translation is half of a primitive translation.

These symmetry elements can be combined in seven different ways, the so-called "seven frieze patterns" (and corresponding groups). In addition to pure translations or translations combined with one of the other four types, we have two additional frieze groups, both containing *translations* and *perpendicular reflections*, combined either with a *parallel reflection* or with a *glide*. In both cases, *rotations* are always present as well. The seven frieze groups are illustrated schematically in Figures 2.1–2.4, and Figures 2.5–2.8 show a series of beautiful Roman mosaics, which contain examples of all seven frieze groups.

Fig. 2.5 A detail of the Megalopsychia mosaic (fifth century AD, Yakto village near Daphne, Turkey). The symmetry is $p211$. From (Hayes, 2011). Photograph by Dick Osserman.

Fig. 2.6 A detail from the border of the Megalopsychia mosaic (fifth century AD, Yakto village near Daphne, Turkey). The symmetry is $p11m$. From (Hayes, 2011). Photograph by Dick Osserman.

Fig. 2.7 A mosaic from the "Tomb of Amerimnia" (Calmness), fourth century Antioch, Turkey, showing different types of frieze symmetry. From the center outwards: $p2mm$, $p1m1$, $p2mg$, $p1m1$. From (Hayes, 2011). Photograph by Dick Osserman.

Fig. 2.8 Part of a splendid "carpet" mosaic, found in an upper level of the "House of the Bird Rinceau" in Daphne and dating from 526–540 AD. The mosaic was divided among sponsoring institutions after excavation; this is known as the Worcester fragment. The symmetry of the bottom frieze is $p11g$. The top frieze has symmetry $p1$, but note that introducing *color* would increase the symmetry of the fragment, since the pattern is symmetric by two-fold rotation combined with black-white interchange (Speiser, 1929; Schubnikov and Belov, 1964). Color symmetry is sometimes used in crystallography to describe magnetic structures (Opechowski and Guccione, 1965). From (Hayes, 2011). (Courtesy of the Worchester Art Museem.)

2.1 Symbols for frieze groups

The new symmetry elements in Figures 2.1–2.4 are shown in a symbolic manner, as in the case of point groups. The symbols for the new symmetry elements are:

- **Translations** are shown both with arrows ($\rightarrow$) and by means of a repeated unit. The choice of the latter, however, is arbitrary, in that we could have chosen a shifted repeated unit or even one with a different shape.
- **Glides** are represented by a dashed bold line, always parallel to the periodic direction.

2.2 Operator graphs generalized to frieze patterns

For point groups, we have defined the *graph* of a given operator as the set of points that are invariant under that particular operator. This seems to present a problem now, since neither translations nor glide leave any point invariant. We can, however, easily generalize the above definitions to glide planes and other operators of the same type, collectively known as *roto-translations*:

- **Operator graphs**. These are *sets of points* in space that are individually (i.e., point-by-point) or collectively (i.e., as a set) invariant (i.e., not moved) upon the application of a given operator. Examples of the former are mirror planes or axes graphs; of the latter glide planes graphs (more to follow in 3D). We draw graphs with special symbols indicating how the operator acts, and we denote the graph of the operator g as $[g]$.

- **Translation graphs**. Based on our definition, the graphs of translations should be sets of points on parallel lines filling the whole space, which is not a convenient representation. For our purposes we will represent translations graphically as "arrows", on the understanding that they do not have an origin, and they can be applied anywhere.
- **Generalized rotations**. Operators that have a point-by-point invariant graph, such as those introduced for the point groups, and including mirror lines, are collectively known as *generalized rotations*. The are called **proper** if they preserve orientation/chirality (as for axes of any order), **improper** if they do not (as for mirrors).
- **Generalized roto-translations**. These are operators that have collectively (but not point-by-point) invariant graphs. They are also classified as proper or improper using the same criterion as above. Simple translations are trivial examples. So far, we have only seen another example, an improper roto-translation (the glide plane).

2.3 More formal concepts for frieze groups

Here, we introduce a few more formal definitions related to the frieze groups, extending analogous concepts already introduced for the point groups.

- **Multiplication tables**. Frieze groups, wallpaper groups and space groups have group order of infinity, and as such should have multiplication tables of size infinity × infinity. However, it is still possible to define a "reduced" multiplication table, provided that we do not insist on complete closure. The most convenient way to achieve this is to include all elements *modulo* a primitive translation. **Note**: As for point groups, the multiplication tables define the *abstract group* in the mathematical sense. There are two frieze groups ($p211$ and $p1m1$) that share the same abstract group structure. The situation where distinct geometrical groups are associated with the same abstract group, which we have seen for point groups and frieze groups, is not encountered for space groups in any dimension – this is known as Bieberbach's theorem (Bieberbach, 1910).
- **Repeat unit or unit cell**. A minimal (but never unique, i.e., always conventional) part of the pattern that generates the whole pattern by application of *pure translations*.
- **Asymmetric unit**. A minimal (but never unique) part of the pattern that generates the whole pattern by application of all the operators. It can be shown that there is always a simply connected choice of asymmetric unit.

Problem 2.2: construct examples of repeat and asymmetric units for the seven frieze groups.

- **Multiplicity**. This is the number of equivalent points *in the unit cell*.
- **Points of special symmetry**. These are points that are invariant by application of one or more operators, and have therefore reduced multiplicity with respect to "general positions". This is analogous to the case of the point groups. They are essentially the graphs of generalized rotations and their intersections. As we shall see, the generalized rotation operators intersecting in each given point define a *point group*, known as the *local symmetry group* for that point.
- **Generators**. As for point groups, these are a set of operators (not necessarily minimal or unique) that generates the whole group by composition. The primitive translation is not required (but is often included) among the generators if a glide is present (Problem 2.1). **Generators are always chosen to belong to different conjugation classes**.

2.4 Relation between graph symmetry and composition

As we have seen for point groups, operator graphs can be considered parts of the pattern. Therefore they are subject to symmetry transformations like everything else. For example, if we have the "initial graphs" in Figure 2.9, then we can generate the other graphs by successive applications of each operator to the *graphs* of the others.

Fig. 2.9 Successive generation of mirror and rotation operators by graph symmetry.

We shall therefore say that all these new operators are generated by **graph symmetry**. More formally, this is equivalent to saying that these operators are generated by *conjugation*, and all belong to the same conjugation class. In particular, we can immediately notice that operators with graphs related by a single primitive translation or a multiple thereof are generated by graph symmetry through that translations, and therefore all belong to the same conjugation class.

2.5 Why did we introduce graphs in the first place? Graphical multiplication tables

The usefulness of graphs stems from the fact that we know intuitively how the operator associated to each graph will act on the pattern, as our brain is very clever at figuring out symmetry. Therefore, generating symmetry-related pattern fragments and new operators by graph symmetry is completely straightforward. Unfortunately, graph symmetry generation does not exhaust the whole multiplication table. Our experience with point groups should have alerted us to the fact that operators related by *composition* (rather than *conjugation*) are not easy to spot. Multiplication tables, even in the reduced form described above, are awkward tools for large groups, and are all but intuitive. What we need is a **graphical multiplication table** and/or a set of rules (such as the ones introduced for the point groups) to construct all the operators. With the help of these rules and tables we will obtain a set of graphs in space plus the unit translations, collectively known as a *group diagram*, which completely defines the group, without the need to introduce coordinates. This is the most natural and elegant way to generate even the most complex three-dimensional space groups.

To avoid confusion, it is important here to pause and stress again what we are trying to accomplish by introducing this graphical notation. We intend to demonstrate graphically how symmetry operators *multiply* by showing how the *graphs* of the operators that are being multiplied and the result of the multiplication are related. Below, we will encounter the simple example of two mirror planes intersecting at 90°. The *composition* of the two reflection operators associated with the two mirror planes is a two-fold rotation, the *graph* of which is the line at the intersection of the two mirror planes. The reader should make sure he or she understands this simple example before proceeding further. To lighten the language, in the remainder we will often use expressions such as "multiply two mirror plane operators" when we really mean "multiply (or compose) the symmetry operators, the graphs of which are the two mirror planes", etc.

2.6 A graphical multiplication table for frieze patterns

A graphical multiplication table is a group diagram fragment, the symmetry elements of which are obtained by multiplying by each other the corresponding symmetry operators. Therefore the symmetry elements of a graphical multiplication table always appear together. In addition to the graph symmetry operations that we already know, only four such fragments are required to define the frieze groups (Figure 2.10). From the figure, it is immediately apparent that **there is a correspondence between graphical multiplication tables and point groups**. Each of the point groups that are relevant for the frieze groups (1, 2, m and $2/m$) has one or more entries in the graphical multiplication tables (the trivial entry for 1 is not shown). In essence, *entries in the graphical multiplication tables describe all the possible ways to combine symmetry elements of the point groups with translations*.

Fig. 2.10 Graphical multiplication table for the frieze patterns.

As we have done before for the point groups, we can also express the same combinations using sets of rules. From left to right, our multiplication tables result from the following rules:

Rule 4 Rotation axes are repeated at a distance of *half* the primitive translation. This results from multiplication of the axis operator 2 with the translation operator t.

Rule 5 For the very same reason, vertical mirror planes .*m*. are repeated at a distance of *half* the primitive translation.

Rule 6 Multiplication of a rotation axis operator 2 with a mirror plane operator .*m*. *not intersecting* it results in a glide plane g.

Rule 7 Another obvious fact (Problem 2.1) resulting from multiplication of ..g with itself is that the "glide" translation is half of a primitive translation. This is because, if g is a glide then $g^2 = g \circ g$ is a translation of twice the length of the glide.

Very important note: Points lying on the *graphs* of the same type of operator are equivalent (i.e., contain the same pattern) *only* if they are related by graph symmetry (i.e., if the two operators are conjugated), *but not* if they result from multiplication. For example, in the leftmost fragment of Figure 2.10, the *middle* axis is *not* related by symmetry (not conjugated) to the other two, so it is not forced to contain the same pattern. Look, for example, at Figure 2.3, or, more interestingly, at Figure 2.5: the patterns around the two-fold axes (the "dot" and the center of the "S" for Figure 2.5) are clearly not the same. The consequence of this is that **there can be several inequivalent points having the same local symmetry**.

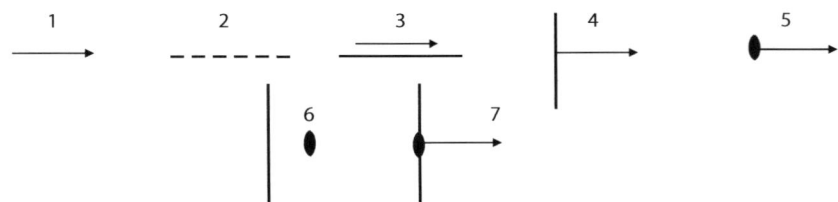

Fig. 2.11 Generators for the frieze groups (not in order).

Problem 2.3: demonstrate graphically the validity of each of the entries in Figure 2.10. Follow the same pattern as for Figure 1.5.

Problem 2.4: construct the seven frieze patterns graphically using the generators in Figure 2.11, the graph symmetry rules and the graphical multiplication tables in Figure 2.10. Can you persuade yourself that there can be no others?

2.7 Commutation: how to "switch" operators

Even in the simple case of frieze patterns, the sequence of application of the operators *does* matter. This is easily seen from the example in Figure 2.12. It is very important to be able to switch operators, and once again, we will do this graphically by means of graph symmetry. Let us have a closer look at Eqn. (1.5). We can reformulate it by introducing the symbol $\tilde{h}_g$, defined as

$$\tilde{h}_g = (g[h]) = g \circ h \circ g^{-1}. \qquad (2.1)$$

In other words, $\tilde{h}_g$ is the operator transformed by graph symmetry through the operator g – i.e., conjugated with h through g. From Eqn. (2.1) the commutation relations between operators follow very naturally:

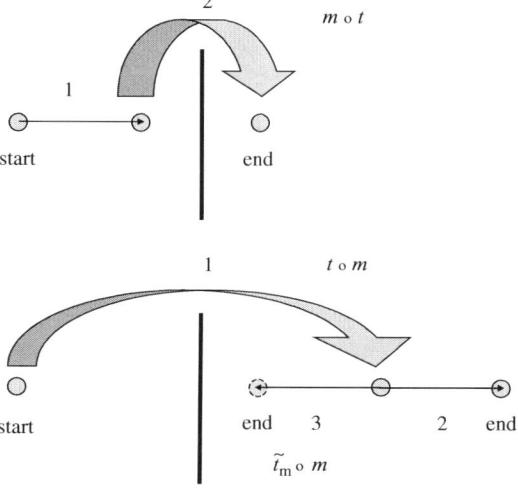

Fig. 2.12 A simple example to show that the order of application of the operators does matter. Applying a translation and then a mirror is not the same as applying the mirror first. To go back to the same point, we would need to apply the operator $\tilde{t}_m$, as explained in the text.

$$g \circ h = \tilde{h}_g \circ g \qquad (2.2)$$
$$h \circ g = g \circ \tilde{h}_{g^{-1}}.$$

We may read this as follows: to pass an operator h from the right to the left of another operator g, we need to transform h by graph symmetry through g (conjugate h through g). As a natural corollary it follows from Eqn. (2.2) that *two operators commute if their graphs are mutually invariant*.

Let us see how this applies to the example in Figure 2.12. Equation (2.2) says that $m \circ t = \tilde{t}_m \circ m$, where, in this case, $\tilde{t}_m$ is the mirror image of the translation, i.e., the translation in the *opposite* direction.

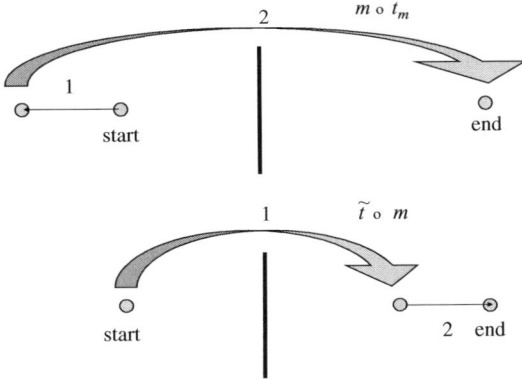

Fig. 2.13 An illustration of how the commutation relations work on the example in Figure 2.12.

2.8 Introducing an origin: normal form for operators

At this point, it is useful for us to introduce the concept of the *origin* of the pattern. We will not introduce a coordinate system yet, but we will single out more or less arbitrarily a point in the pattern for further use. For reasons that will become clear shortly it is useful (although by no means necessary) for this point to be halfway between the top and the bottom of the frieze pattern. Our aim is now to write every symmetry operator h in the so-called *normal* form

$$h = t \circ r \qquad (2.3)$$

where t is a translation (not necessarily a *symmetry* translation) and r is a proper or improper rotation (*not* a roto-translation) that *intersects the origin*. Two important points: first, r and t are *not* in general symmetry operators; second, the order of the operators matters: r is applied first, followed by t.

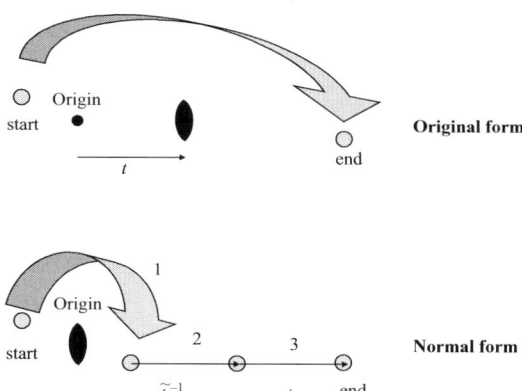

Fig. 2.14 An illustration of how an off-origin axis is converted to normal form.

We can distinguish three cases:

1. **Horizontal symmetry elements**, i.e., ..*m* and ..*g* for frieze patterns, already intersect the origin (provided that the origin is chosen halfway between the top and the bottom of the frieze pattern). These operators are therefore *already in normal form* (remember that the glide $g = t \circ (..m)$).
2. **Vertical mirrors and axes through the origin** are already in normal form, and need no further discussion.
3. **Vertical mirrors and axes *not* passing through the origin**. It is easy to see that the *graph* of these operators is simply the *translated graph* of an operator of case 2 (in general the translation in not a symmetry operator). They can therefore be written (see Eqn. 1.5) as $h = t \circ r \circ t^{-1}$ or, using Eqn. (2.2),

$$h = t \circ \widetilde{t_r^{-1}} \circ r \qquad (2.4)$$

which is of course in normal form, since the composition of two translations is a translation. Note that both rotations and vertical mirrors reverse the direction of the translations, so $t \circ \widetilde{t_r^{-1}} = 2t$. This procedure is illustrated in Figure 2.14.

The conclusion is that **all symmetry operators can be converted to normal form**. We will see that this is valid more generally for wallpaper and space groups. Clearly, the normal form of each operator will be different depending on the choice of origin. However, the *rotational* part never changes (as expressed with respect to its own origin).

2.9 Choice of origin

As already mentioned, the origin can be chosen in an arbitrary way. However, some choices are more convenient than others. In particular, it is useful to choose a **point of special symmetry** as the origin, so that as many operators as possible are already in normal form.

Problem 2.5: look at the frieze patterns in Figures 2.1–2.4. Can you suggest good choices for the origins in each case?

2.10 Composition of operators in normal form

Using the commutation relations in Eqn. (2.2), it is straightforward to establish how to compose two operators in normal form to yield another operator in normal form:

$$(t \circ r) \circ (t' \circ r') = (t + \widetilde{t}'_r) \circ (r \circ r') \tag{2.5}$$

where we have used the "+" sign to indicate explicitly that composition of translations is the simple *vector sum*.

2.11 Point groups and local symmetry

Equation (2.5) shows something extremely important: **the rotational parts of the normal-form operators of a group form *themselves a group*.** This new group is much simpler than the original one, because it does not contain any translation, and is therefore *one of the point groups we already introduced*. This is known as the *crystal class* of the frieze pattern. Crystal classes were first introduced by the Finnish-born Russian artillery professor and amateur crystallographer Alex Gadolin (1828–1892) (Gadolin, 1867). Naturally, not all the point groups we already know can represent crystal classes for the frieze groups. Since only two-fold axes are allowed, the only frieze crystal classes are 1, 2, *m* and 2*mm*.

Problem 2.6: for each of the frieze groups in Figures 2.1–2.4, construct the corresponding crystal classes (point groups.)

It can be easily shown that the set of *generalized rotations* intersecting at a given point is also a point group – the point being the fixed point of the point group. This group is known as the *local symmetry group* for that point.

2.12 Coordinate systems for frieze groups

We will once again defer the full discussion of crystallographic coordinate systems. However, for the limited purpose of understanding the ITC entries for the frieze groups, we will introduce an *ad hoc*, extremely simple coordinate system. This is a Cartesian system, with the x-axis

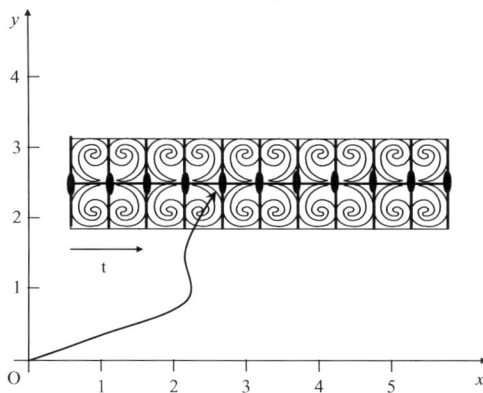

Fig. 2.15 Coordinate system for frieze patterns, illustrated here for the case of the 2*mm* group. The origin is to be translated to one of the points of special symmetry of the pattern (curvy line).

along the repetition direction and the *y*-axis orthogonal to it. The unit length (in both directions) coincides with the length of the primitive translations. This is illustrated in Figure 2.15

2.13 Frieze groups in the ITC

The seven frieze groups are listed in ITC, Volume E (Kopský and Litvin, 2002) on pages 30–36. The entry for the frieze group *p2mg* is shown in Figure 2.16. We can now understand most of the entries in these tables.

- **First line**. From left to right, the entries are for the frieze group, the crystal class and the crystal system. The frieze group symbol (known as the Hermann–Mauguin symbol) contains four characters. The first is always a *p*, and indicates that *primitive* translations are symmetry elements. The second symbol (either 1 or 2) indicates the absence or presence of a two-fold rotation. The third symbol (1 or *m*) indicates the presence or absence of a mirror line *orthogonal to the repeat direction*. The fourth symbol (1 or *m* or *g*) refers to the symmetry elements *parallel* to the repeat direction.
- **Second line**. From left to right, the entries are a sequence of numbers from 1 to 7, a repetition of the Hermann–Mauguin symbol (for space groups, this entry contains an "extended" symbol) and the "Patterson symmetry" – see Giacovazzo *et al.*, (2002) for a discussion of Patterson symmetry.
- **Diagrams**. Two diagrams are shown: the left-hand diagram shows the arrangement of the symmetry elements within one *unit cell*; the right-hand one shows a general position and its equivalents, also within one unit cell. By longstanding crystallographic convention (and contrary to everyone else), **the *a*-axis points vertically**

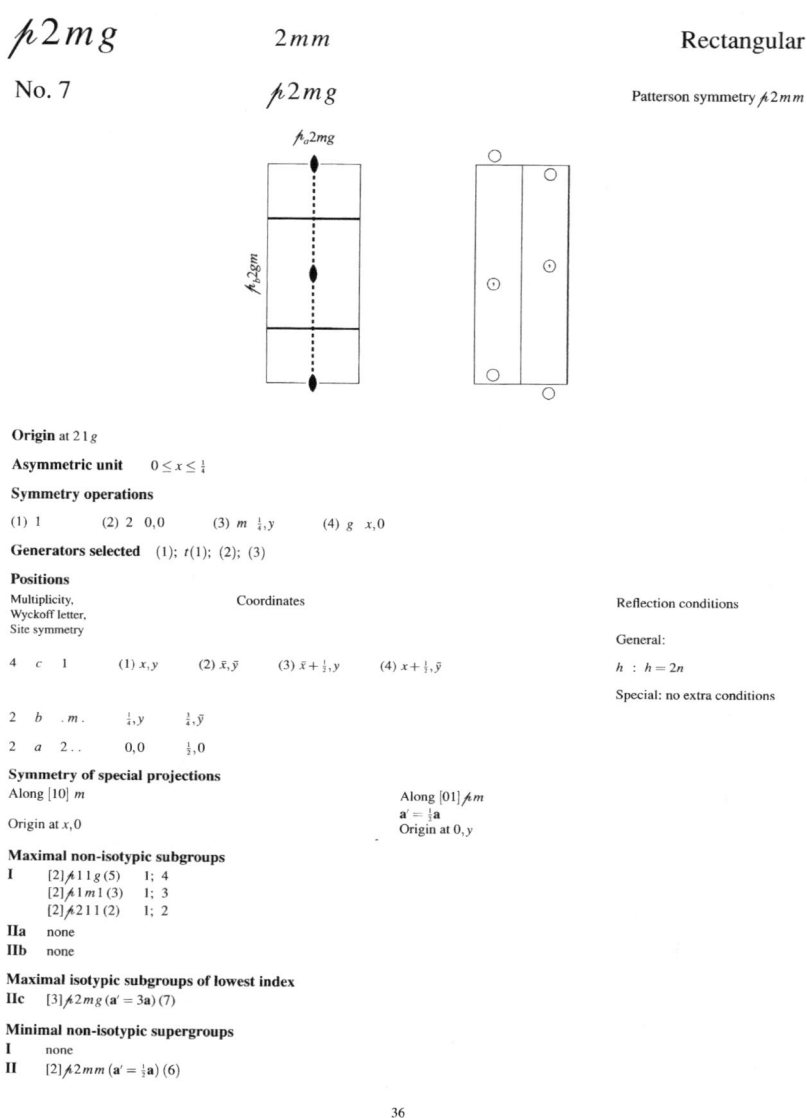

Fig. 2.16 The frieze group *p2mg* from ITC, Volume E, page 36 (Kopský and Litvin, 2002).

downwards, whereas the *b*-axis points to the right. The axes and the unit cell are chosen to be *symmetric by the crystal class*. In the right-hand diagram, general points are represented with circles. Circles with a comma (",") are related by an odd number of reflections to circles without the comma. This is of course immaterial if they are to represent points, but it may matter if we were to "dress" the points with attributes such as a polar vector or a chiral molecule. In these diagrams, the *origin* (see next line) is at the center of the diagram. The diagram can be rotated to give a different "setting" of the frieze group, which differs simply by the axes conventions. **Note**: frieze groups belonging to the "oblique" systems (1 and 2) are shown with a non-orthogonal set of axes and an oblique unit cell. Although this conforms to symmetry, one could equally well adopt Cartesian coordinates, since the y-direction is non-periodic (see below). We have adopted a simpler orthogonal system in Figure 2.1.

- **Statement of the origin**. The origin chosen for the subsequent entries is stated here.
- **Asymmetric unit**. One choice of the asymmetric unit is given here. All the positions listed below are within the *primitive* unit cell, provided that the first point x, y, z is within the *asymmetric* unit cell.
- **Symmetry operators**. All the inequivalent symmetry operators within the asymmetric unit cell (excluding the translation) are listed here. The operators are not listed in normal form. Rather, the type of symmetry operator is listed, followed by a position within the asymmetric unit cell that uniquely locates the symmetry element. For example, the entry (3) $m \frac{1}{4}, y$ in Figure 2.16 indicates the presence of a mirror plane at $\frac{1}{4}$ along the x-direction and parallel to y.
- **Generator selected**. A set of generators for the frieze group, not necessarily minimal. The first generator is always 1, the second is the primitive translation t. The others are chosen from the symmetry operators given above.
- **Positions**. The general and special positions for the frieze group. The entries for Columns 1–3 are the same as for the point groups.

    Column 4 Coordinates. A general position and its equivalent positions is listed first. The equivalent positions are obtained by applying the symmetry operators listed above in the listed order. For the higher-symmetry positions, the same order is followed but identical positions are omitted.

    Column 5 Reflection conditions. We will defer the discussion of this entry.

- **Symmetry of special projections**. This entry indicates the symmetry of projections of the patters along a (a point group) or b (a 1D line group).
- **Subgroups and supergroups**. A list of maximal subgroups and minimal supergroups follows. A complex classification scheme, which we will not describe in detail, is used to generate this list. Note that "isotypic" subgroups have the same Hermann–Mauguin symbol but different periodicity (larger unit cell). For example, the entry $[2]p11g(5)$ 1;4 has the following meaning: The subgroup has index 2 ($[2]$), has H–M symbol $p11g$, corresponds

to frieze group number 5 and has symmetry operators 1 and 4 in the list above. An entry such as ($a' = 3a$) means that the subgroup has tripled periodicity with respect to the original group. The study of subgroups and their classification was initiated by Carl Hermann (1929).

Wallpaper (plane) groups 3

Wallpaper groups describe the symmetry of patterns that are repetitive in two dimensions. In the case of true wallpapers, the repetition vectors tend to be orthogonal, because the process of hanging the wallpaper usually involves lining up identical patterns on straight horizontal lines. However, no such restriction applies, for example, to textiles, pavements or other decorative forms in two dimensions. Once again, we will ignore the possibility of color symmetry (i.e., the application of a symmetry operator combined with exchange or permutation of colors). In the previous sections, we have already introduced all the necessary symmetry operators, namely in the form of two-dimensional point groups. However, the system of translations required to describe the wallpaper groups is significantly more complex than the simple, one-dimensional system used for frieze groups. We will therefore devote some space to the topic of translations in 2D, with particular reference to the symmetry of the translation set.

No new operators need to be introduced to describe the wallpaper groups, since combinations of the point-group and frieze-group operators are all that is required. The composition rules are the same as before, and can be worked out graphically. The most significant new issue is the introduction of *lattices*. As we shall see shortly, lattices are a convenient way of representing the set of all symmetry translations, which is *always* a subgroup of the planar and space groups.

3.1 Transformation of translations by graph symmetry

The first step in constructing the lattice is to consider which translation operators may be equivalent by symmetry, i.e., belong to the same conjugation class. It is clear that pure translations cannot be conjugated through other translations, since all translations commute (see also Problem 1.6). Therefore, we need to consider how a translation operator t is transformed by *graph symmetry* by another operator g that is *not* a pure translation. We will employ conjugation as the "formal" definition of graph symmetry, as given in Eqn. (1.5). We will write the operator g^{-1} in *normal form* as $g^{-1} = s \circ r^{-1}$, s being a translation, and write g

in *reverse form* $g = r \circ -s$. It is easily seen that these two operators are the inverse of each other. Considering that translations commute, we obtain:

$$g \circ t \circ g^{-1} = r \circ -s \circ t \circ s \circ r^{-1} = r \circ t \circ r^{-1}. \tag{3.1}$$

In other words, *when transforming translations by graph symmetry, only the rotational part, i.e., only the operator of the point group, needs to be considered*. Graphically, we can represent the graph of the operators with arrows, as we have done before, and keep the "tail" as the fixed point of the point group. All the arrows that can be generated by applying the operators of the point group, i.e., those defining the *crystal class*, will be equivalent by graph symmetry, and must therefore all be present as symmetry operators.

3.2 The symmetry of the translation set

From the previous paragraphs, one might get the impression that the symmetry of the translation set must be the same as the crystal class. However, this is not so. The symmetry is *at least* as high as that of the crystal class, but it can be higher. For example, in 2D, the symmetry of the translation set must necessarily contain the two-fold axis, since for each translation t, $-t$ is also present (in 3D, the *center of symmetry* must be present). In some cases, the symmetry of the translations is even higher. The point group representing the symmetry of the translation system is known as its *holohedry*.

3.3 Lattices

Lattices are an alternative representation of the translation set. These are sets of points generated from a single point (the *origin*) by applying all the translation operators, and can be thought of as representing the graphs of all the translation operators. Once the origin is chosen, the translations uniquely define the lattice. Conversely, all the translations can be obtained as position vectors of each lattice point with respect to the origin.

3.4 Crystallographic restriction

In order to construct groups that contain translations as well as other operators, we need, for each generalized rotation, to generate a set of translations that are "compatible" with each other. Not all finite groups are compatible with a discrete lattice; in any dimension, we will have only a finite number of compatible groups. This is the essential reason why only 10 *crystallographic* point groups exist in 2D, and only four can be employed for the frieze patterns. As we shall see, the *angles*

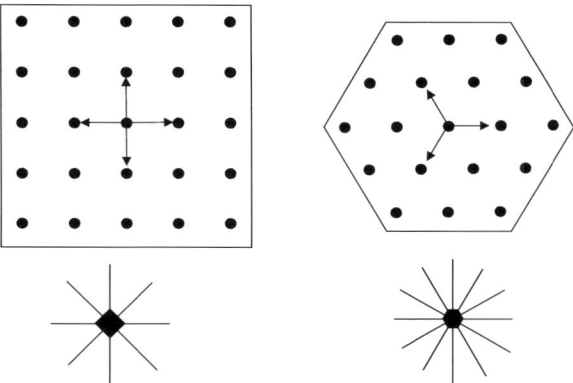

Fig. 3.1 Portions of the square and hexagonal lattices, with their respective point symmetry groups. Note that the symmetry of the lattice is higher than that of the minimal point group needed to construct them from a single translation (4 and 3, respectively)

between the translation vectors and the symmetry elements can also be restricted.

We can easily construct lattices (or equivalently, sets of translations) that are symmetric after two-, three-, four- and six-fold rotations (Figure 3.1). Why are axes of other orders not allowed? The proof that no lattice can be constructed to support axes of different orders is contained in an elegant theorem, proven *ex absurdo*, known as the **restriction theorem**. In essence, one postulates that a given symmetry translation is the *shortest* one in the translation set (this is always possible, since infinitesimal translations are disallowed). One then constructs the symmetry-equivalent translations, and proves that, by taking linear combinations of those, one can construct a translation *shorter* than the original one.

Figure 3.2 shown a graphical representation of the restriction theorem, showing that axes of order higher than six are not compatible with *any* 2D lattice. This construction does not apply to five-fold symmetry, which requires a different but related construction (Figure 3.3).

3.4.1 Restriction on the orientation of symmetry elements on the lattice

A different form of the restriction theorem defines incompatibilities between certain 2D point groups and certain directions of the symmetry translations. This is important, because **it constrains the orientation of the point group on the lattice**. For example, Figure 3.4 shows that in general no minimal translation can form an angle smaller than 30° with a mirror plane unless it is on the plane itself. For point groups 4*mm* and 6*mm* the minimal (shortest) translations must be parallel to the mirror planes. Figure 3.4 also shows that there are always translations parallel and perpendicular to a mirror plane, although neither are necessarily minimal.

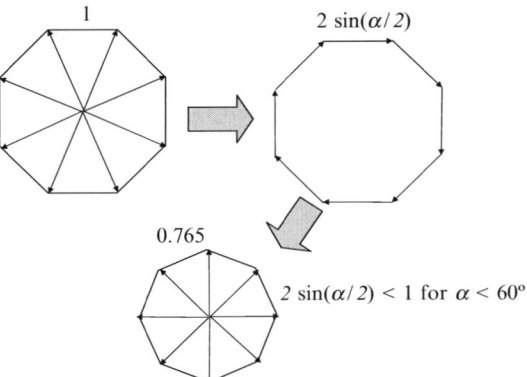

Fig. 3.2 The restriction theorem for $n > 6$. The *differences* between symmetry-equivalent translation vectors form a *restricted* polygon with the same symmetry. Carrying on this process one would obtain arbitrarily small translation vectors, which is contrary to having assumed only discrete symmetry. This can easily be proven also *ex absurdo*, by postulating that the generating translation is the smallest of the set.

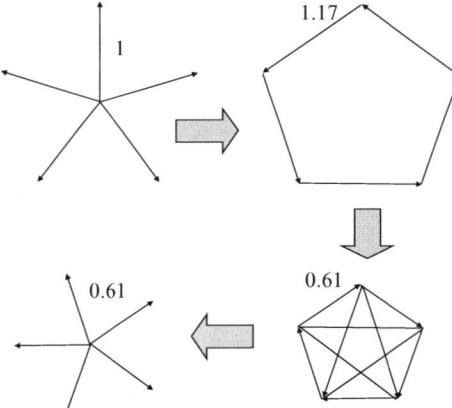

Fig. 3.3 The restriction theorem for $n = 5$. The restricted pentagon is constructing using differences of differences.

3.5 Bravais lattices in 2D

Bravais lattices, named after the French physicist Auguste Bravais (1811–1863) (Bravais, 1850; 1851), define all the translation sets that are mutually compatible with crystallographic point groups. There are five of them: "oblique", "*p*-rectangular", "*c*-rectangular", "square" and "hexagonal". They can all be generated constructively in a simple way.

3.5.1 Oblique system (holohedry 2)

Here, each translation t is symmetry-related only to $-t$, so there is no restriction on the length or orientation of the translations. The resulting

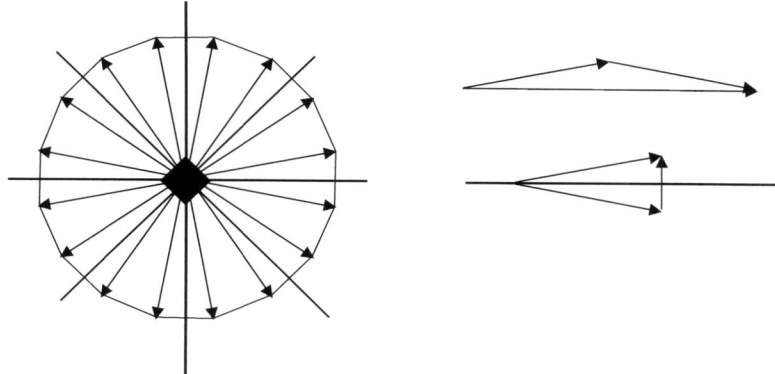

Fig. 3.4 The second form of the restriction theorem applied to point groups containing mirror planes. Left: the shortest translation in the translation set cannot make an angle smaller that 30° with a mirror plane, unless it is on the plane itself. In fact, the *difference* between two such translations would be smaller than the original one. For 4*mm* and 6*mm* symmetries, where the angles between mirror planes are 45° and 30°, this forces the translations to lie on the mirror planes themselves. Right: translations *parallel* and *perpendicular* to the mirror planes always exist as part of the translation set, although *minimal* translations are not restricted in this way other than for the above cases.

lattice is a tiling of parallelograms, and supports the crystal classes 1 and 2.

3.5.2 Rectangular system (holohedry 2*mm*)

Here we have two lattices (Figure 3.5), both supporting crystal classes *m* and 2*mm*:

- Both the shortest translation and the next-to-shortest one not collinear with the first lie on the mirror planes. In this case, the result is simple tiling of rectangles, known as a "*p*-rectangular" (primitive rectangular) lattice.
- Either the shortest or the next-to-shortest non-collinear translation is at an angle with the planes (in the latter case, one can show *by restriction* that its projection on the plane must bisect the shortest translation). The result is a rectangular lattice with nodes at the centers of the rectangles, known as a "*c*-rectangular" (centered rectangular) lattice.

3.5.3 Square system (holohedry 4*mm*)

There are two classes in this system: 4 and 4*mm*. They are both supported by a simple square lattice. In the latter case, as we have already shown, the primitive translations must lie on the mirror planes (Figure 3.1).

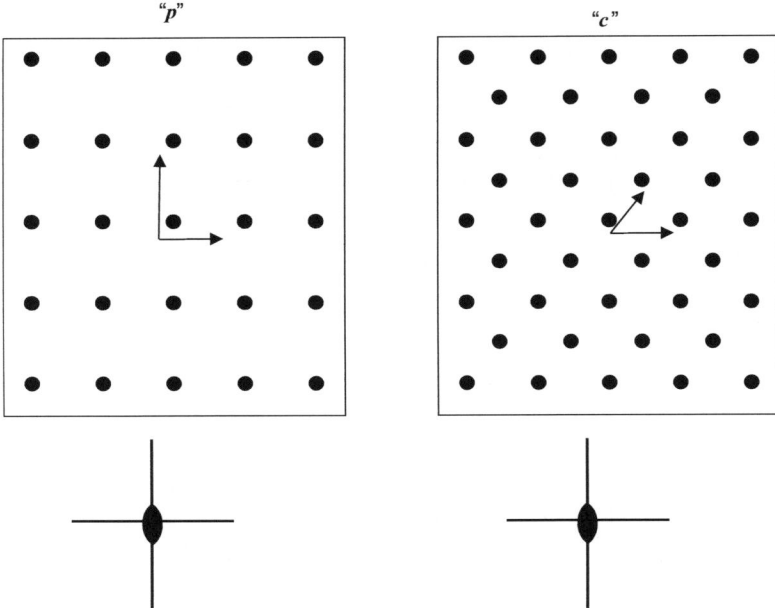

Fig. 3.5 The two types of rectangular lattices ("p" and "c") and their construction.

3.5.4 Hexagonal system (holohedry 6*mm*)

There are four classes in this system: 3, 3*m*1 (or 31*m*), 6 and 6*mm*. They all generate simple hexagonal lattices. In the case of 6*mm*, the nodes must lie on the mirror planes (Figure 3.1), whereas in the case of 31*m* they must by restriction lie either on the mirror planes (setting 31*m*) or exactly in between (setting 3*m*1). Note that here the distinction is real, and will give rise to two different wallpaper groups.

3.6 Unit cells in 2D

We have already introduced the concepts of *primitive* and *asymmetric* unit cell for the case of frieze patterns. These concepts are essentially the same for wallpaper groups, representing minimal units that can generate the whole pattern by translation and by application of all symmetry operators, respectively. It should be noted that a variety of choices are possible for the unit cell, including cells with curvilinear sides, as long as they tile perfectly and have the same areas (Figure 3.6). In particular, one can show that any translation vector that is not a multiple of another can serve as one of the sides of a parallelogram-shaped unit cell. Nevertheless, the natural choice for the *primitive* unit cell, and the one that is usually adopted, is a parallelogram defined by the two linearly independent shortest translations. In the case of the *c*-rectangular lattice, this unit cell is either a rhombus or a parallelogram. The latter does

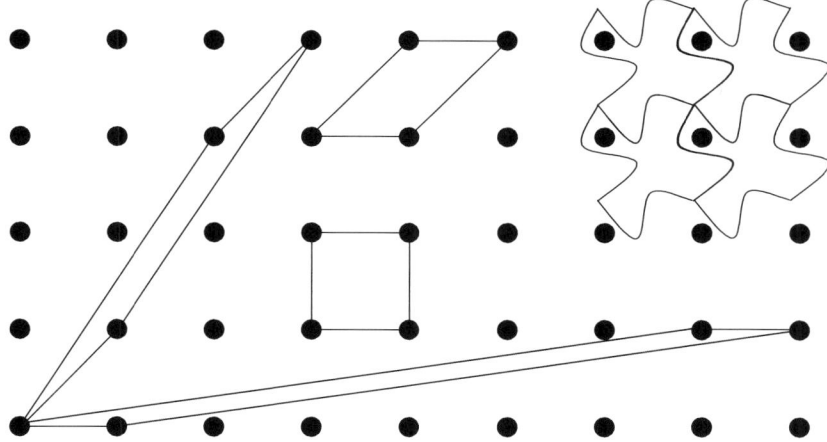

Fig. 3.6 Possible choices for the primitive unit cell on a square lattice.

not possess the full symmetry of the lattice (holohedry), and neither is it particularly convenient to define coordinates (see below). It is therefore customary to introduce a so-called **conventional** or **centered** rectangular unit cell, which has double the area of the primitive unit cell (i.e., it always contains two lattice points), but has the full symmetry of the lattice and is defined by orthogonal translation vectors (Figure 3.7).

3.7 Composition rules in 2D

The last step to construct the wallpaper groups is to determine the composition rules between the allowed operators. Clearly, the rules we have previously established for the point and frieze groups will still be valid, but, with wallpaper groups, more possibilities arise. Now we have axes of order 3, 4 and 6, which can be composed with translations. In addition, translations can be composed with mirror and glide planes at different angles, not only orthogonal or parallel to them. Finally, axes of allowed orders can be composed with mirror planes and glides; the axes can be either on or off the planes. Some of the rules we will derive will be generalizations of the old rules, others will be new. At the end, as in the case of the frieze patterns, we will provide a *graphical* multiplication table to aid the construction of the wallpaper groups.

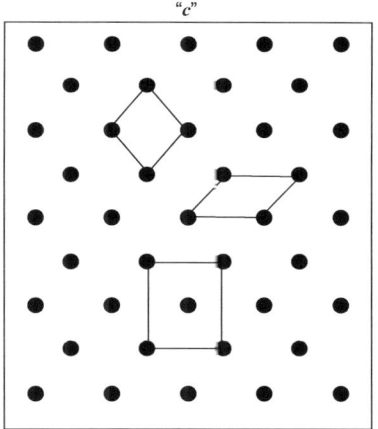

Fig. 3.7 Two primitive cells and the conventional unit cell on a c-centered rectangular lattice.

3.7.1 Composition of translations with axes

In general, the composition of an axis of whatever order with a translation perpendicular to it is *a translated axis of the same order and sense of rotation*. This is straightforward to prove (we leave this to the reader) by taking the n-th power of the new operator (n being the order of the axis) and using the permutation rule. The only thing that needs to be found

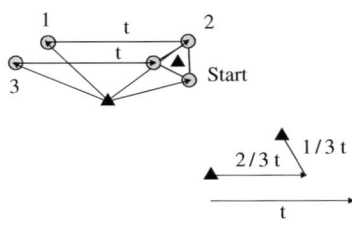

Fig. 3.8 Graphical construction illustrating the composition of a three-fold axis with a translation orthogonal to it.

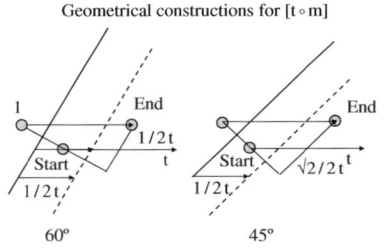

Fig. 3.9 Graphical construction illustrating the composition of mirrors and glides with a translation at 60° and 45° inclination.

is the position of the new axis graph – a very easy task if one exploits the fact that the graph must be invariant. This results in an extension of Rule 4 above:

Rule 4' The composition of a translation t with an orthogonal n-fold rotation axis through an angle $\alpha = 2\pi/n$ (in the order $t \circ r$) is a new rotation axis of the same order, displaced by $t/2$ *along* the translation and "up" (perpendicular to the translation and in the direction defined by the same sense of the rotation) by the amount $\frac{t}{2}\cot\frac{\alpha}{2}$. Two-fold rotation axes are repeated at a distance of *half* the primitive translation. This results from multiplication of the axis operator 2 with the translation operator t.

Problem 3.1: prove Rule 4' above. A specific example for the case $n = 3$ is provided in Figure 3.8

3.7.2 Composition of translations with mirrors and glides

Compositions of translations with mirrors or glides at arbitrary inclinations are easily constructed by decomposing the translation into components parallel and perpendicular to the plane.

Rule 5' The composition of a translation t with a mirror or glide (in the order $t \circ m$ or $t \circ g$) is either a mirror or a glide, displaced by $t/2$ along t and with an added in-plane translation component added to it (making mirrors into glides and vice versa), by an amount corresponding to the in-plane projection of the translation.

Problem 3.2: prove Rule 5' above. A specific example for the cases encountered in wallpaper groups is provided in Figure 3.9

3.7.3 Composition of rotations with mirrors and glides

The graphical construction of the individual compositions is somewhat more involved than the previous ones. Figure 3.10 provides two examples. The resulting rules, however, can be summarized in an extremely simple way.

Rule 8 Mirrors and glides transform rotations parallel to them in their (displaced) reverse operator by graph symmetry. The compositions of a rotation axis with a mirror or glide give rise to a star of planes and glides with the same "shape" as the point group corresponding to the rotation plus mirrors (2mm, 3m1 or 4mm, 6mm). The center of the star is found at the intersection between

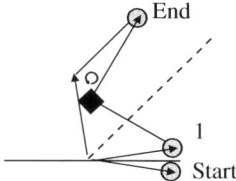

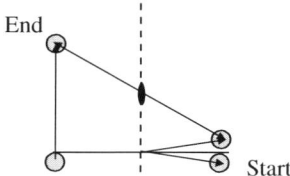

Fig. 3.10 Two examples of composition of mirror planes with parallel rotation axes not lying on them.

the mirror (or glide) and the line joining the rotation graph with its mirror or glide image. The arms of the stars will be mirrors or glides, with an appropriate glide translation (in general, *different for each*), so that they all relate the rotation pair by graph symmetry.

3.8 Graphical multiplication table in 2D

Following the example we set for the frieze pattern, in Figure 3.11 we provide a graphical multiplication table to be used to construct wallpaper groups. As we shall see below, all the groups can be constructed by combining and transforming the fragments in the table and the five Bravais lattices by graph symmetry alone (i.e., we will no longer have to worry about composition). **As in the case of the frieze groups, all the entries in the table are related to point groups, and represent all the possible ways to combine generalized rotations (the elements of point groups) with translations.** One can easily recognize that some of the fragments simply *are* the 2D point groups, whereas others are generated using the aforementioned rules.

3.9 Nomenclature for wallpaper groups

The nomenclature for wallpaper patterns follows the familiar scheme already employed for point groups and frieze groups. The first symbol indicates the lattice, the second the rotation axis, and the third and fourth mirrors or glides along two symmetry-inequivalent directions

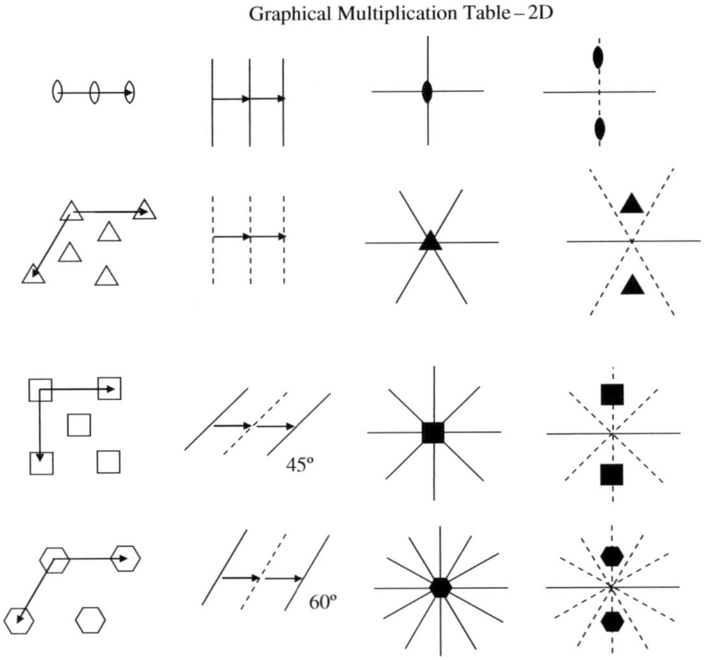

Fig. 3.11 The graphical multiplication table in 2D.

(known as the "secondary" and "tertiary" directions). The only significant novelty is that the first letter will be a "*p*" or "*c*" depending on whether the lattice is primitive or centered. Possible ambiguities are resolved by the following two conventions.

- As we can see from Figure 3.11, there will be cases where parallel mirror and glide planes will be present simultaneously. Here, the convention ("rule of priority") is that *m* takes precedence, so the operator *g* is listed only if there is no *m* parallel to it. Therefore, for example, there is no *m* in *p*2*gg*. Conversely, *cm* contains glide planes, but they are not listed in the symbol since they are parallel to *m*.
- For square and hexagonal lattices (e.g., *p*3*m*1), the *third* symbol is *perpendicular to the lattice translations* ("secondary" symmetry direction), whereas the *fourth* is *perpendicular* to the other ("tertiary") non-equivalent direction (*at* 45° *for the square* and *at* 30° *for the hexagonal*).

3.10 The 17 wallpaper groups

We can construct all the possible candidate wallpaper groups by simply combining the five Bravais lattices with the ten 2D crystal classes, and systematically replace the *m*'s with *g*'s at all locations. This

Table 3.1 The 17 wallpaper groups. The symbols are obtained by combining the five Bravais lattices with the ten 2D point groups, and replacing g with m systematically. Strikeout symbols are forbidden by the "rule of priority" convention, because a mirror *parallel* to g would also be present. $p2mg$ and $p2gm$ are equivalent by (arbitrary) 90° rotation (see text).

Crystal system	Crystal class	Wallpaper groups
Oblique	1	$p1$
	2	$p2$
Rectangular	m	$pm, cm, pg, \text{\sout{cg}}$
	$2mm$	$p2mm, p2mg \,(=p2gm), p2gg, c2mm, \text{\sout{c2mg}}, \text{\sout{c2gg}}$
Square	4	$p4$
	$4mm$	$p4mm, p4gm, \text{\sout{p4mg}}$
Hexagonal	3	$p3$
	$3m1\text{--}31m$	$p3m1, \text{\sout{p3mg}}, p31m, \text{\sout{p31g}}$
	6	$p6$
	$6mm$	$p6mm, \text{\sout{p6mg}}, \text{\sout{p6gm}}, \text{\sout{p6gg}}$

procedure yields 27 symbols (Table 3.1). In some cases, however, we have constructed a symbol whereby a lattice translation is intersecting a glide plane at an angle different from 90°. This is easily seen in the symbol cg, where the "centering" translation is at an angle with the symmetry direction. As we can see from the multiplication table (Figure 3.11), this situation generates a mirror m *parallel* to g, so, by the rule of priority, we should have written cm. The same is true for all the other strikeout operators. The wallpaper groups $p2mg$ and $p2gm$ are equivalent, because both secondary and tertiary symmetry directions are edges of the rectangle, and are chosen arbitrarily. This leaves only 17 valid symbols, corresponding to the 17 wallpaper groups.

3.11 Analyzing wallpaper and other 2D art using wallpaper groups

The symmetry of a given 2D pattern can be readily analyzed and assigned to one of the wallpaper groups, using one of several schemes. One should be careful in relying too much on the lattice symmetry, since it can often be higher than the underlying pattern (especially for true wallpapers). Mirrors and axes are quite easily identified, although, once again, one should be careful with pseudo-symmetries. Figure 3.12 shows a decision-making diagram that can assist in the identification of the wallpaper group (a similar flow-chart-based diagram was first proposed by Cotton, 1990). Here, no reliance is made on the lattice, although sometimes centering is easier to identify than glides. Figures 3.13–3.18 show a few 2D patterns from various sources, with the associated wallpaper group. In the caption, the rationale for the choice is explained. Many more examples are available in the cited sources.

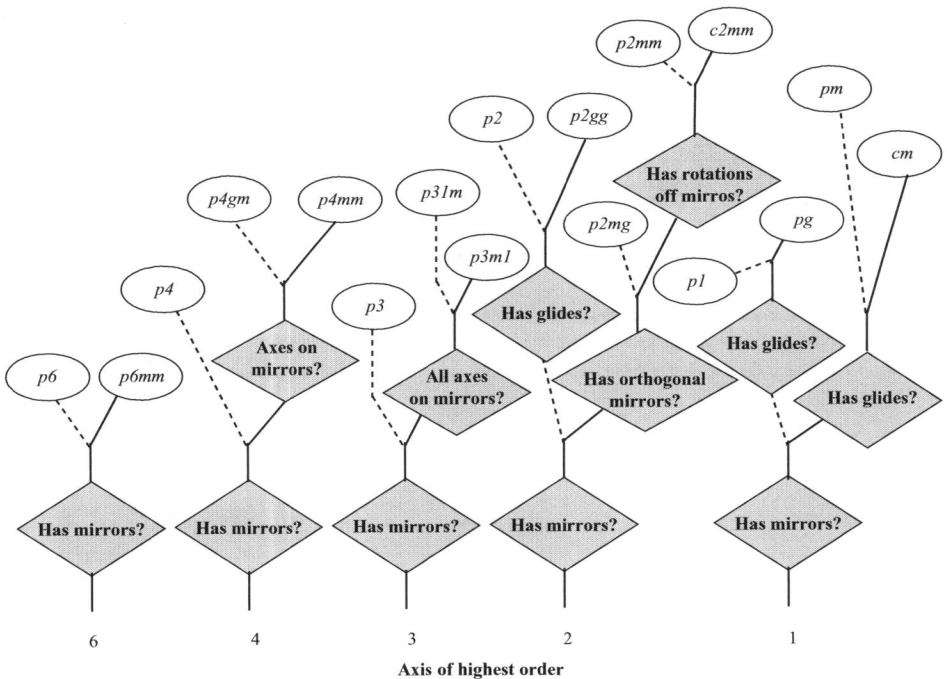

Fig. 3.12 Decision-making tree to identify wallpaper patterns. The first step (bottom) is to identify the axis of highest order. Continuous and dotted lines are "Yes" and "No" branches, respectively. Diamonds are branching points.

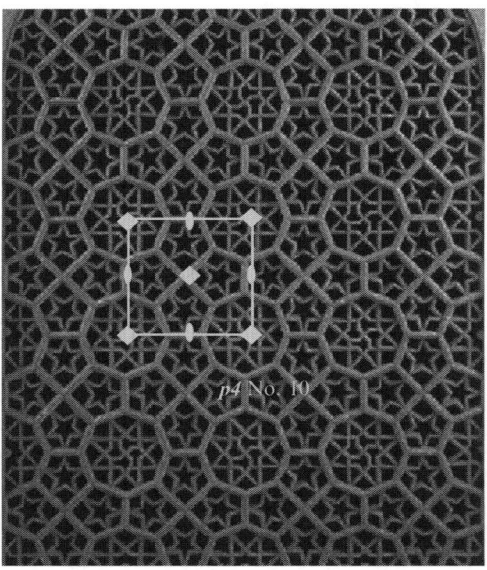

Fig. 3.13 Jali pierced screen (one of a pair), second half of the 16th century; Mughal, probably from Fatehpur Sikri, India, carved red sandstone (Jali, 2000). Metropolitan Museum of Art, cat. no. (1993.67.2). The highest-order rotation is 4. The four-armed hooked crosses inside the octagons all turn in the same direction, so there cannot be mirror planes. The wallpaper group is therefore $p4$.

Analyzing wallpaper and other 2D art using wallpaper groups 47

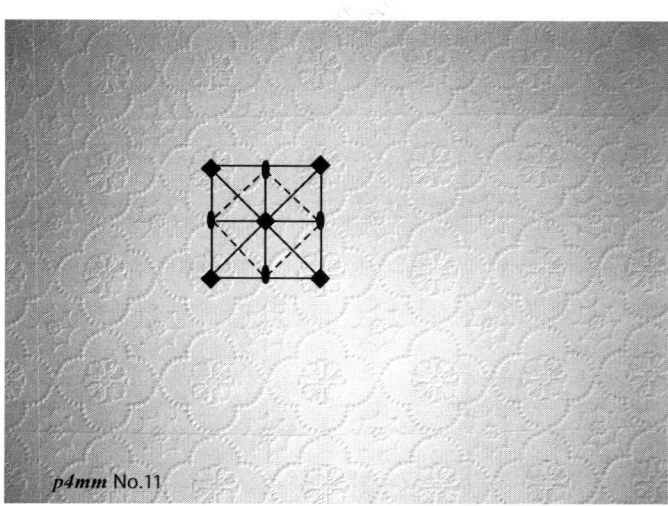

Fig. 3.14 A pattern from the ceiling of the author's home. The highest-order rotation is 4, and there are mirror planes on the four-fold axes (two inequivalent ones). The symmetry is *p*4*mm*.

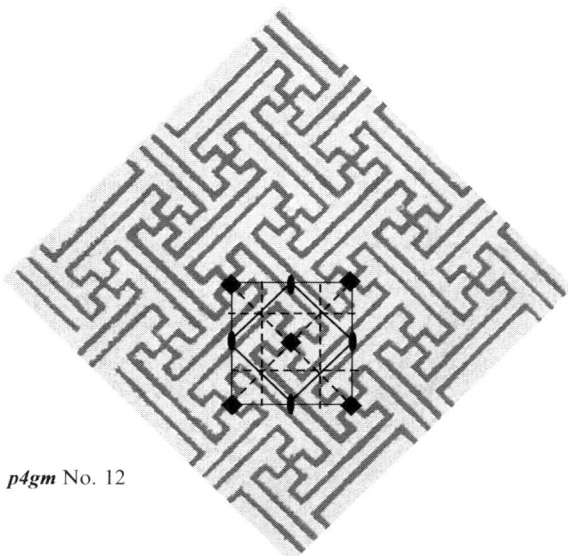

Fig. 3.15 A Chinese pattern from Owen (1865). The highest-order rotation is 4, and there are mirror planes relating the hooked crosses, but the four-fold axes are off them. The space group is *p*4*gm*.

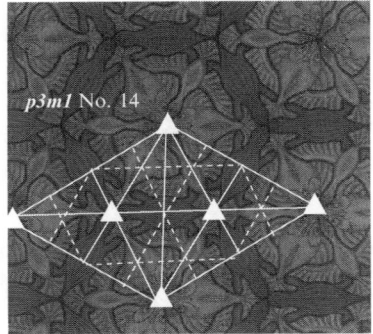

Fig. 3.16 Escher drawing of fishes and turtles (Schattschneider and Hofstadter, 2004). There are three types of three-fold sites (the heads of the fishes and the heads and tails of the turtles), both with mirror symmetry. The group is *p*3*m*1.

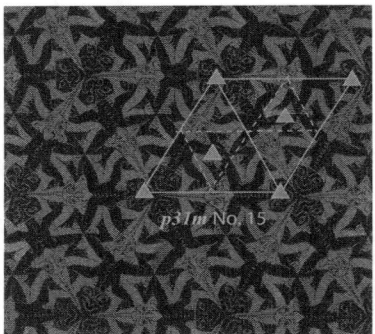

Fig. 3.17 Escher drawing of devils (Schattschneider and Hofstadter, 2004). Mirror symmetry is present, but only on the heads of the devils, not on their hands. The group is *p*31*m*.

3.12 Constructing wallpaper groups

Sadly, being able to construct space groups from their Hermann–Mauguin symbol is no longer counted among the essential skills of a crystallographer, and is indeed a Herculean task for some of the most complex cubic space groups. It is a skill that can, however, be mastered readily and completely for the planar groups – a good exercise that will also enable one to construct a large number of relatively simple space groups. Here, we provide a recipe to construct all the 17 wallpaper groups, and work out in detail one of the most complex examples (*p*6*mm*).

1. Draw the two mirror symmetry elements corresponding to the secondary and tertiary symmetry directions at an appropriate angle (90° for the rectangular lattice, 45° for the square lattice and 30° for the hexagonal lattice). If one or both symbols are missing (or contain a 1), go to the next step.
2. Define a provisional origin at the intersection of the two symmetry elements (or arbitrarily if no mirror is present; in this case the rotation axis is to be placed at the origin). Locate the rotation axis to be consistent with these symmetry elements and with one of the fragments in the 2D graphical multiplication table (Figure 3.11). The six-fold axis and the first three-fold axis to be located always

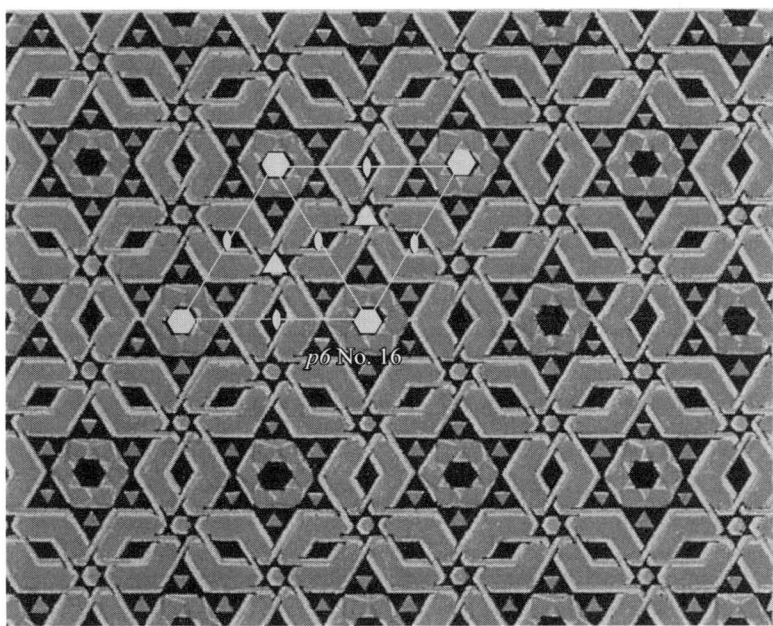

Fig. 3.18 An Egyptian pattern from Owen (1865). The hexagons have six-fold symmetry, while the hooked crosses have only three-fold (in spite of appearances) All rotate clockwise, so there cannot be any mirror. Group *p*6.

go on the mirror plane, if present. At this point, one may want to redefine the origin to be on a rotation axis, if it is not already there.
3. Define the translation vectors as "arrows", and sketch the lattice. Be careful with "c" lattices. All the translation vectors must lie on a symmetry direction (i.e., perpendicular to a plane or glide), except for the case of *p*31*m*, where they are parallel to the mirrors.
4. Propagate the fragment by graph symmetry with the translations (i.e., replicate the fragment along all the translation vectors). Join mirror lines as appropriate. You can draw the conventional unit cell at this point.
5. Propagate the rotations and mirrors by *composition* with the translations (e.g., repeat all the two-fold axes at half periodicity). This needs to be repeated also for the implicit axes contained in axes of higher order (2 inside 4, 2 and 3 inside 6). Join mirror lines as appropriate.
6. Look for axes of any order related by a mirror *m*, and "decorate" the mid-point with glides according to the fragments in Figure 3.11. Mirrors and glides may need to be repeated in a pattern consistent with non-orthogonal translations.

An example of how to construct the rather complex wallpaper group *p*6*mm* is given in Figure 3.19.

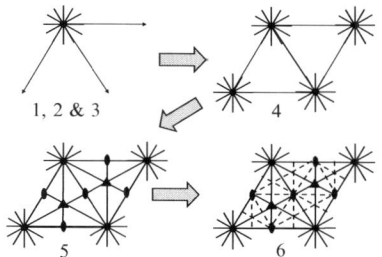

Fig. 3.19 Construction of the wallpaper group *p6mm* using the graphical multiplication table in 2D. The numbers below each diagram indicate the steps followed up to that point (see text). Note that in step 6, the horizontal glides (and equivalent) arise from six-fold axes separated by a mirror. The vertical glides (and equivalent) arise from two-fold or three-fold axes separated by a mirror.

CRYSTALLOGRAPHIC COORDINATE SYSTEMS

Part II

Up to this point, we have relied mostly on the graphical representation of symmetry operators, which is completely free from conventions, as long as the operator graphs have been located on a given pattern. For a full understanding of the entries in the ITC, and, perhaps more importantly, for their practical use, it is necessary at this point to introduce crystallographic coordinate systems and the mathematical representation of the symmetry operators that they engender. Crystallographic coordinate systems are a subset of the possible coordinate systems in a Euclidean space, a vast topic that in itself is distinct from crystallography. However, as we shall see, only coordinate systems with basis vectors coinciding with translation operators are employed in crystallography. We will also discuss measurements of distances and angles in generalized coordinates, employing the tensor notation, and, in particular, through the use of the metric tensor. Finally, we will introduce the so-called dual basis, which will later serve to construct the reciprocal lattice and to define the Fourier transforms of lattice functions. Although we have only examined symmetry groups up to 2D, we will treat these topics in 3D, as there is no significant difficulty introduced by the extra dimension. In Chapter 7, we will combine the graphical representation of symmetry elements with the knowledge of coordinate systems to describe a phase transition in two dimensions.

Coordinate systems in crystallography

4

Before we begin discussing the different coordinate systems employed in crystallography, it is natural to address the most basic question a reader may have: why don't we generally employ Cartesian coordinates in crystallography? The answer is that in crystallography we almost exclusively employ coordinate systems with **basis vectors coinciding with either primitive or conventional translation operators**. When primitive translations are used as basis vectors, **points of the pattern related by translation will differ by** *integral values of x, y and z*. **When conventional translations are used as basis vectors, points of the pattern related by translation will differ by either** *integral* **or** *simple fractional (either $n/2$ or $n/3$ values of x, y and z)*. This advantage **far outweighs** the convenience of using Cartesian coordinates. In this chapter, we will introduce a simple notation that will enable the reader to deal with the most complex settings and to transform between them "almost" effortlessly.

4.0.1 Coordinates and components

It is important to remark on the distinction between **points** of the space and **vectors**. **Points** of a Euclidean space form a so-called **affine space**, and are themselves not vectors, because the "sum" of two points and the "multiplication" of a point by a scalar are not defined. Rather, an affine space has an auxiliary vector space "attached" to it, over which these operations are defined; *"differences"* between points are uniquely associated with vectors in this auxiliary space. We may then write

$$p_2 - p_1 = \mathbf{v} \tag{4.1}$$

and its inverse, i.e., the *sum* of a point with a vector, yielding another point

$$p_2 = p_1 + \mathbf{v}. \tag{4.2}$$

For Euclidean spaces, the dot product is also defined, so that one can define the length ("norm") of vectors and the angles between them. Through the previous definitions, this translates into the ability of measuring **distances between points** (the norm of their difference) and **angles between segments**.

Once an *origin point* "o" and a *basis for the vector space* are chosen, the *coordinates* of a point p are the *components* of the difference vector $p - o$. This special difference vector is known as the **position vector**. A *coordinate transformation* therefore may involve two things: (a) a *change of origin* and (b) a *change of basis* for the vector space.

Naturally, once a choice of basis has been made, that basis can be used to express vectors *other* than position vectors in terms of their *components*.

Note: In crystallography, *basis vectors* have the dimension of a *length*, and *coordinates* (position vector components) are *dimensionless*.

We will first describe **generalized** transformations of *basis vectors and components*, and then include **changes of origin** to describe **generalized** *coordinate* transformations

4.0.2 Notation

To describe vectors in terms of their *components*, we will employ a special but very convenient "tensor" notation. We will denote the basis vectors as $\mathbf{a}_i$ (note the *subscript*), where the correspondence with the usual crystallographic notation is

$$\mathbf{a}_1 = \mathbf{a}; \quad \mathbf{a}_2 = \mathbf{b}; \quad \mathbf{a}_3 = \mathbf{c}. \tag{4.3}$$

We will sometimes employ explicit array and matrix multiplication for clarity. In this case, the array of basis vectors is written as a *row*, as in $[\mathbf{a}] = [\mathbf{a}_1 \ \mathbf{a}_2 \ \mathbf{a}_3]$. The corresponding *column* array will be denoted as $[\mathbf{a}]^T$.

Components of a generic vector $\mathbf{v}$ will be denoted as v^i (note the *superscript*), where

$$v^1 = v_x; \quad v^2 = v_y; \quad v^3 = v_z. \tag{4.4}$$

Components will be expressed using *column* arrays, as in $[v] = \begin{bmatrix} v^1 \\ v^2 \\ v^3 \end{bmatrix}$,

whereas the *row* vector will be denoted by $[v]^T$.

Components of a **position vector** will be denoted as $x^i = \begin{bmatrix} x \\ y \\ z \end{bmatrix}$.

4.0.3 Basis vector and component transformations – covariant and contravariant quantities

We now want to describe the effect of changing between two systems of basis vectors, say $\mathbf{a}_i$ and $\mathbf{a}'_i$. The new components can be expressed in term of the old ones, as

$$[\mathbf{a}'_1, \mathbf{a}'_2, \mathbf{a}'_3] = [\mathbf{a}_1, \mathbf{a}_2, \mathbf{a}_3]\mathbf{P} \tag{4.5}$$

$$= [\mathbf{a}_1, \mathbf{a}_2, \mathbf{a}_3] \begin{bmatrix} P_{11} & P_{12} & P_{13} \\ P_{21} & P_{22} & P_{23} \\ P_{31} & P_{32} & P_{33} \end{bmatrix}$$

or, in a more compact form, $\mathbf{a}'_j = \mathbf{a}_i P^i_j$.

In the tensor notation, indices with the same letter appearing in the subscript and superscript are implicitly summed (we sometimes use the word "contracted indices" to indicate this implicit summation):

$$A^i_j B^j_k \equiv \sum_j A^i_j B^j_k. \tag{4.6}$$

Likewise, we can write the corresponding coordinate transformations as

$$\begin{bmatrix} v'^1 \\ v'^2 \\ v'^3 \end{bmatrix} = \mathbf{Q} \begin{bmatrix} v^1 \\ v^2 \\ v^3 \end{bmatrix} \tag{4.7}$$

$$= \begin{bmatrix} Q_{11} & Q_{12} & Q_{13} \\ Q_{21} & Q_{22} & Q_{23} \\ Q_{31} & Q_{32} & Q_{33} \end{bmatrix} \begin{bmatrix} v^1 \\ v^2 \\ v^3 \end{bmatrix}$$

or, in a more compact form, $v'^i = Q^i_j v^j$.

We now observe that the vector $\mathbf{v}$ is an *invariant* quantity – in other words, we are simply describing two ways of expressing the same vector $\mathbf{v}$, which in itself **does not depend on the choice of coordinates**. Therefore

$$\mathbf{v} = \mathbf{a}_i v^i = \mathbf{a}'_j v'^j = \mathbf{a}_i P^i_j Q^j_k v^k \tag{4.8}$$

whence

$$P^i_j Q^j_k = \delta^i_k \rightarrow \mathbf{Q} = \mathbf{P}^{-1}. \tag{4.9}$$

In other words, the matrices $\mathbf{Q}$ and $\mathbf{P}$ are inverse of each other.

Quantities that transform like the basis vectors are called *covariant* and written with **subscript indices**, whereas quantities that transform like the components are called *contravariant* and written with **superscript indices**. The notation is chosen in such a way that when *all* the superscript and subscript indices are "contracted" (summed), the resulting quantity is **invariant**.

Problem 4.1: determine the matrices $\mathbf{P}$ and $\mathbf{Q}$ for a coordinate transformation from a hexagonal lattice to the corresponding C-centered rectangular lattice.

4.1 Change of origin and generalized coordinate transformations

As discussed above a general *coordinate* transformation entails a change in both origin and basis vectors. If a new origin o' is chosen, the point p will be associated with a new "position" vector

$$\mathbf{v}' = p - o' = p - o + o - o' = \mathbf{v} - \mathbf{s} \qquad (4.10)$$

where

$$\mathbf{s} = o' - o. \qquad (4.11)$$

The vector **s** has its "head" on the old origin and its "tail" on the new origin. In order to find the new coordinates, we will therefore have to determine the components of the vector **v** in terms of the new basis vectors, and add to them the components of the "origin shift" vector $\mathbf{s} = o' - o$, also expressed in terms of the new basis vectors.

4.2 The most general coordinate transformation

By employing Eqns (4.10) and (4.7) we are now ready to write the most general transformation of coordinates for a point p in the affine space. We recall that the coordinates of p are simply the components of the vector $\mathbf{v} = p - o$, so as such they must be written in contravariant form as x^i. Likewise, the components s^i of the origin shift vector **s** are also contravariant. We will therefore write

$$x'^i = Q^i_j x^j - s'^i = Q^i_j (x^j - s^j). \qquad (4.12)$$

We stress again that the s^j are the components of the new origin with respect to the old origin, expressed using the old basis vectors.

The inverse equation is

$$x^i = (Q^{-1})^i_j (x'^j + s'^j) = (Q^{-1})^i_j x'^j + s^j. \qquad (4.13)$$

The mathematical form of symmetry operators

5

5.1 Symmetry operators in Cartesian coordinates

We shall begin by making the simplest assumption on the coordinate system – that of Cartesian coordinates – and derive the form of all the symmetry operators. Pseudo-Cartesian coordinates are employed in crystallography for square (or cubic) symmetries, and also for lower symmetries, provided that the translation vectors (either primitive or those defining the edges of the centered cell) are accidentally equal in length and orthogonal to each other – this type of lattice is described as *metrically* square or cubic. Note that, unlike the case of "true" Cartesian coordinates, *crystallographic* coordinates are *always dimensionless* (they are known as *fractional coordinates*), so the basis vectors have the dimension of a *length*.

We already know from previous sections that all operators can be written in normal form with respect to an arbitrary origin, which we will choose to coincide with the origin of the coordinate system. In Cartesian coordinates, a translation of point p_1 into point p_2 by the operator t is simply the *sum* of the coordinates of p_1 (say, x_1, y_1, z_1) and the Cartesian components of the translation vector (say, t_x, t_y, t_z). Since the translation components are common to all points, they do not affect either distances or angles between different points, since these are calculated on the *vectors* connecting the pairs of points.

In Cartesian coordinates, a rotation of a point about an axis through the origin is represented by an *orthogonal matrix* R. This is easy to see, for instance, by requiring that the distance d_2 between the new point and the origin is the same as that from the old point and the origin (d_1):

$$\begin{bmatrix} x_2 \\ y_2 \\ z_2 \end{bmatrix} = R \begin{bmatrix} x_1 \\ y_1 \\ z_1 \end{bmatrix} \qquad (5.1)$$

$$(d_2)^2 = [x_2 \ y_2 \ z_2] \begin{bmatrix} x_2 \\ y_2 \\ z_2 \end{bmatrix} = [x_1 \ y_1 \ z_1] \mathbf{R}^T \mathbf{R} \begin{bmatrix} x_1 \\ y_1 \\ z_1 \end{bmatrix} \quad (5.2)$$

$$= [x_1 \ y_1 \ z_1] \begin{bmatrix} x_1 \\ y_1 \\ z_1 \end{bmatrix} = (d_1)^2.$$

Equation (5.2) implies $\mathbf{R}^T \mathbf{R} = \mathbb{1}$, i.e., by definition, $\mathbf{R}$ is an orthogonal matrix:

$$\mathbf{R}^T = \mathbf{R}^{-1}. \quad (5.3)$$

One can easily see that $\det(\mathbf{R}) = \pm 1$, due to the properties of the matrix determinant.

$$\det(\mathbf{R}^T) = \det(\mathbf{R}) \quad (5.4)$$

$$\det(\mathbf{R}^{-1}) = \det(\mathbf{R})^{-1}.$$

Proper rotations have $\det(\mathbf{R}) = 1$, whereas *improper* rotations have $\det(\mathbf{R}) = -1$.

In three dimensions (or, more generally, in the case of an odd number of dimensions), proper and improper rotations can be related to each other through the *inversion* operator I, in itself an improper rotation, with matrix representation $\mathbb{I} = -\mathbb{1}$. In other words, for each proper rotation there is an improper one obtained by changing the sign of all coordinates. **Note**: what is improper in 3D may be proper in 2D. For example, the inversion in 2D coincides with the two-fold rotation, and is therefore proper.

By combining the translational and rotational parts of normal-form operators, we find the general expression of all symmetry operators in Cartesian coordinates:

$$\begin{bmatrix} x_2 \\ y_2 \\ z_2 \end{bmatrix} = \begin{bmatrix} t_x \\ t_y \\ t_z \end{bmatrix} + \mathbf{R} \begin{bmatrix} x_1 \\ y_1 \\ z_1 \end{bmatrix}. \quad (5.5)$$

5.2 Rotation matrices

It is easy to deduce the form of all possible proper and improper rotation matrices in 3D. Let us start by writing in matrix form rotation through an angle θ around each coordinate axis:

$$\mathbf{R}_x = \begin{bmatrix} 1 & 0 & 0 \\ 0 & \cos\theta & -\sin\theta \\ 0 & \sin\theta & \cos\theta \end{bmatrix} \quad (5.6)$$

$$R_y = \begin{bmatrix} \cos\theta & 0 & \sin\theta \\ 0 & 1 & 0 \\ -\sin\theta & 0 & \cos\theta \end{bmatrix}$$

$$R_z = \begin{bmatrix} \cos\theta & -\sin\theta & 0 \\ \sin\theta & \cos\theta & 0 \\ 0 & 0 & 1 \end{bmatrix}.$$

Rotations around arbitrary axes can be thought of as transformations of the above rotations by *graph symmetry*. For example, a rotation around the [1, 1, 0] direction (between the x and y axes) can be thought of as a rotation of 45° around z of the *graph* of the rotation around the x-axis. Now, the rule of transformation by graph symmetry becomes really useful:

$$R_{[1,1,0]} = R_z(45°) \cdot R_x \cdot R_z(-45°). \tag{5.7}$$

Problem 5.1: determine the matrix form of the "cubic" rotation "3", i.e., a rotation of 120° around the diagonal of the cube, assuming that the coordinate axes run along the cube edges.

Another useful formula provides the matrix for a rotation through an angle θ around an axis with arbitrary orientation, given by the unit vector x, y, z:

$$\mathcal{M}(\hat{v}, \theta) = \begin{bmatrix} \cos\theta + (1-\cos\theta)x^2 & (1-\cos\theta)xy - (\sin\theta)z & (1-\cos\theta)xz + (\sin\theta)y \\ (1-\cos\theta)yx + (\sin\theta)z & \cos\theta + (1-\cos\theta)y^2 & (1-\cos\theta)yz - (\sin\theta)x \\ (1-\cos\theta)zx - (\sin\theta)y & (1-\cos\theta)zy + (\sin\theta)x & \cos\theta + (1-\cos\theta)z^2 \end{bmatrix}. \tag{5.8}$$

As already mentioned, the corresponding *improper* rotations can be obtained by multiplication by the matrix $\mathbb{I}$ (note that $\mathbb{I}$ commutes with all operators). The crystallographic notation for this is a *bar* above the rotation symbol, as in $\bar{3}$. It is noteworthy that mirror planes result from the composition (multiplication) of two-fold axes *orthogonal* to them with inversion, so that, for example, $\bar{2}_z = m_z$. We will examine later on the meaning of the other improper axes $\bar{3}, \bar{4}$ and $\bar{6}$.

5.3 Symmetry operators in generalized crystallographic coordinates

Having determined the general form of a symmetry operator in Cartesian coordinates and the most general way to transform those coordinates to any other system, we can easily write the general form of a symmetry operator in *any* coordinate system. We will first rewrite Eqn. (5.5) using tensor notation. However, to avoid confusion between the dimensional indices and the point label, we will write the latter in parentheses after the symbol:

$$x^i(2) = t^i + R^i_j x^j(1) \quad \text{or, in matrix form} \quad [x(2)] = [t] + \boldsymbol{R}[x(1)]. \tag{5.9}$$

Next, we multiply by $\boldsymbol{Q}$ and $\boldsymbol{Q}^{-1}$, as

$$\boldsymbol{Q}[x(2)] = \boldsymbol{Q}[t] + \boldsymbol{Q}\boldsymbol{R}\boldsymbol{Q}^{-1}\boldsymbol{Q}[x(1)]. \tag{5.10}$$

Finally, we employ Eqn. (4.12):

$$[x'(2)] + \boldsymbol{Q}[s] = \boldsymbol{Q}[t] + \boldsymbol{Q}\boldsymbol{R}\boldsymbol{Q}^{-1}([x'(1)] + \boldsymbol{Q}[s]) \tag{5.11}$$

or

$$[x'(2)] = \boldsymbol{Q}([t] - [s] + \boldsymbol{R}[s]) + \boldsymbol{Q}\boldsymbol{R}\boldsymbol{Q}^{-1}[x'(1)]$$

or, in component form

$$x'(2)^i = Q^i_j(t^j - s^j - R^j_k s^k) + Q^i_k R^k_l (Q^{-1})^l_j x'(1)^j.$$

Note that Eqns (5.9) and (5.11) have exactly the same form. The only difference is that the rotational part $\boldsymbol{Q}\boldsymbol{R}\boldsymbol{Q}^{-1}$ is *not* a unitary matrix, unless, of course, $\boldsymbol{Q}$ itself is unitary, meaning that the coordinate transformation is a roto-translation of the axes, which would therefore still remain Cartesian.

In three dimensions, $\det(\boldsymbol{Q}\boldsymbol{R}\boldsymbol{Q}^{-1}) = \pm 1$. In the former case, it describes what is known as a **proper rotation**; in the latter an **improper rotation**, such as a reflection or inversion. **Improper** rotations are operations that change the **handedness** (left- to right-hand or vice versa). It can be shown that **all improper rotations can be obtained by composing a proper rotation with inversion**.

Problem 5.2: determine the matrix form of the symmetry operator representing a rotation by 60° in hexagonal coordinates.

5.3.1 Coordinate transformations and symmetry operators: avoid the confusion!

At this point, one can observe that Eqns (5.5) and (4.12) are also of very similar form. In fact, one may be tempted to interpret Eqn. (5.5) as a special kind of coordinate transformation. However, the two concepts must remain fully distinct to avoid confusion. Based on our initial convention of treating all symmetry transformations as *active*, symmetry operators *always* relate two distinct points p_1 and p_2, whereas coordinate transformations *always* refer to a *single* point, which is described using different coordinates.

Distances, angles and the real and reciprocal spaces

6

6.1 Determination of distances and angles in real space: the metric tensor

In Cartesian coordinates, the scalar product between two vectors takes the familiar form

$$\mathbf{v} \cdot \mathbf{u} = [v]^T [u] = a^2 \delta_{ij} v^i u^j \qquad (6.1)$$

where $a^2 = |\mathbf{a}|$ is the length of the basis vector in whatever units (with dimensions) are employed (it is generally 1 for Cartesian coordinates with a dimensionless basis). We shall now see how Eqn. (6.1) can be generalized to *crystallographic* coordinates, which are in general non-Cartesian and have basis vectors with the dimension of a *length*. Remembering the generic expression of $\mathbf{v}$ and $\mathbf{u}$ we can write

$$\mathbf{v} \cdot \mathbf{u} = \mathbf{a}_i u^i \cdot \mathbf{a}_j v^j = [\mathbf{a}_i \cdot \mathbf{a}_j] u^i v^j. \qquad (6.2)$$

The quantity in square brackets is **symmetric**

$$G_{ij} = \mathbf{a}_i \cdot \mathbf{a}_j \qquad (6.3)$$

and is known as the **metric tensor**; it has the dimension of a *length squared*. The metric tensor G_{ij} enables one to calculate the dot product in *any* coordinate system, as

$$\mathbf{v} \cdot \mathbf{u} = G_{ij} v^i u^j. \qquad (6.4)$$

Once the dot product is known, one can easily determine distances and angles between points in any coordinate system. Here is how the metric tensor can be constructed given the lattice parameters, and how distances and angles are calculated:

- One is generally given the **lattice parameters** $\mathbf{a}$, $\mathbf{b}$, $\mathbf{c}$, α, β and γ. In terms of these, the metric tensor can be written as

$$G = \begin{bmatrix} a^2 & ab\cos\gamma & ac\cos\beta \\ ab\cos\gamma & b^2 & bc\cos\alpha \\ ac\cos\beta & bc\cos\alpha & c^2 \end{bmatrix}. \quad (6.5)$$

- To measure the **length** v of a vector **v**:

$$v^2 = |\mathbf{v}|^2 = \begin{bmatrix} v^1 & v^2 & v^3 \end{bmatrix} G \begin{bmatrix} v^1 \\ v^2 \\ v^3 \end{bmatrix}. \quad (6.6)$$

- To measure the **angle** θ between two vectors **v** and **u**:

$$\cos\theta = \frac{1}{uv} \begin{bmatrix} u^1 & u^2 & u^3 \end{bmatrix} G \begin{bmatrix} v^1 \\ v^2 \\ v^3 \end{bmatrix}. \quad (6.7)$$

Problem 6.1: write the explicit form of the dot product between two vectors in the most general case (triclinic).

6.2 Dual basis and coordinates: the reciprocal space

Let us assume a basis vector set $\mathbf{a}_i$ for our vector space as before, and let us consider the following set of *new* vectors:[1]

$$\mathbf{b}^i = 2\pi \mathbf{a}_k (G^{-1})^{ki}. \quad (6.8)$$

From Eqn. (6.3) it follows that:

$$\mathbf{a}_i \cdot \mathbf{b}^j = 2\pi \mathbf{a}_i \cdot \mathbf{a}_k (G^{-1})^{kj} = 2\pi G_{ik}(G^{-1})^{kj} = 2\pi \delta_i^j. \quad (6.9)$$

Note that the vectors $\mathbf{b}^i$ have dimensions $length^{-1}$. The following properties can be easily proven:

1. The vectors $\mathbf{b}^i$, considered as a set, are *contravariant*. This could be shown explicitly, but it is clear from the position of the indices.
2. The vectors $\mathbf{b}^i$ are **linearly independent** since the $\mathbf{a}_i$ are. One can therefore use the $\mathbf{b}^i$ as new basis vectors, forming the so-called ***dual basis***. This being a perfectly legitimate choice, we can express any vector on this new basis as

$$\mathbf{q} = q_i \mathbf{b}^i \quad (6.10)$$

where the coordinates q_i are necessarily *covariant*. As we have just said, we can write any vector in this new basis, but vectors expressed using **dimensionless coordinates** in the dual basis have dimensions $length^{-1}$, and cannot therefore be summed to the position vectors. We can consider these vectors as *representing the position vectors of a separate space*, the so-called ***reciprocal space***.

---

[1] Note that the factor of 2π in Eqn. (6.8) is omitted in many crystallographic textbooks. Those of you familiar with the notation used in general relativity may object to the notation of Eqn. (6.8), which emphasizes the fact that the *covariant* and *contravariant* metric tensor *matrices* (see also below) are inverse of each other (except for the factor of 2π). In the notation used in general relativity, this is implicitly assumed, and the factor of 2π is omitted. Therefore the *covariant* metric tensor g_{ij} and the *contravariant* metric tensor g^{jk} are related by $g_{ij}g^{jk} = \delta_i^k$.

Position vectors in reciprocal space are linear combinations of the dual basis vectors with *dimensionless* components. Their dimension is $length^{-1}$. If a *primitive basis* is used for "direct" (normal) space, then *reciprocal lattice vectors* are reciprocal-space position vectors with *integral components*.

3. The dot product between position vectors in real and reciprocal space is a dimensionless quantity, and has an extremely simple form:

$$\mathbf{q} \cdot \mathbf{v} = 2\pi q_i x^i. \qquad (6.11)$$

4. In particular, the dot product of integral multiples of the original basis vectors (i.e., **direct** or **real lattice vectors**), with integral multiples of the dual basis vectors (i.e., **reciprocal lattice vectors**) are integral multiples of 2π. This property will be used extensively to calculate Fourier transforms of lattice functions.

5. The reciprocal-space metric tensor is $\tilde{\mathbf{G}} = (2\pi)^2 \mathbf{G}^{-1}$, so that, for two reciprocal-space vectors $\mathbf{q}$ and $\mathbf{s}$:

$$\mathbf{q} \cdot \mathbf{s} = \tilde{G}^{ij} q_i s_j. \qquad (6.12)$$

The length of a vector in reciprocal space is a key concept in crystallography, as it is related to the spacing between crystal planes (known as the d-spacing), as:

$$d = \frac{2\pi}{|q|}. \qquad (6.13)$$

The classic book by Warren (1990), p. 20, provides the *spacing formulas* for all crystal systems.

6.2.1 Orientation of reciprocal space

In considering reciprocal space as distinct from direct space, one should not forget that its basis has been obtained by coordinate transformation from the direct (real-space) basis, through Eqn. (6.8). Therefore, nothing stops us expressing real-space vectors using the dual basis and vice versa, so long as we are prepared to use non-dimensionless components. In particular, it is possible to determine the *orientation* of a reciprocal lattice vector (or tensor) with respect to a real-space vector, by simply taking the dot product of the normalized vectors. Another common application of this type of transformation is the determination of the atomic displacement ellipsoids, expressed as quadratic forms in *real space*, from the Debye–Waller tensors, expressed as quadratic forms in *reciprocal space*. The rank-2 atomic displacement tensors (ellipsoids) define the *real-space* amplitude and direction of the harmonic atomic vibrations, and are directly proportional to the Debye–Waller tensors, obtained from X-ray or neutron diffraction experiments by modeling the scattered intensity along different directions in reciprocal space.

6.2.2 Recapitulation of the key formulas for the dual basis

- From **direct** to **dual** bases (Eqn. 6.8)

$$\mathbf{b}^i = 2\pi \mathbf{a}_k (G^{-1})^{ki}.$$

- Dot product relation between the two bases (Eqn. 6.9)

$$\mathbf{a}_i \cdot \mathbf{b}^j = 2\pi \delta_i^j.$$

- Dot product between vectors expressed in the two *different* bases (Eqn. 6.11)

$$\mathbf{q} \cdot \mathbf{v} = 2\pi q_i x^i.$$

- Dot products between vectors in *real space* (Eqn. 6.4)

$$\mathbf{v} \cdot \mathbf{u} = G_{ij} v^i u^j. \qquad (6.14)$$

- Dot products between vectors in *reciprocal space* (Eqn. 6.12)

$$\mathbf{q} \cdot \mathbf{r} = \tilde{G}^{ij} q_i r_j. \qquad (6.15)$$

6.2.3 Dual basis in 3D

In three dimensions, there is a very useful formula to calculate the dual basis vectors, which makes use of the properties of the vector product:

$$\mathbf{b}_1 = 2\pi \frac{\mathbf{a}_2 \times \mathbf{a}_3}{\mathbf{a}_1 \cdot (\mathbf{a}_2 \times \mathbf{a}_3)} \qquad (6.16)$$

$$\mathbf{b}_2 = 2\pi \frac{\mathbf{a}_3 \times \mathbf{a}_1}{\mathbf{a}_1 \cdot (\mathbf{a}_2 \times \mathbf{a}_3)}$$

$$\mathbf{b}_3 = 2\pi \frac{\mathbf{a}_1 \times \mathbf{a}_2}{\mathbf{a}_1 \cdot (\mathbf{a}_2 \times \mathbf{a}_3)}.$$

Note that

$$v = \mathbf{a}_1 \cdot (\mathbf{a}_2 \times \mathbf{a}_3) = abc \left(1 - \cos^2 \alpha - \cos^2 \beta - \cos^2 \gamma + 2 \cos \alpha \cos \beta \cos \gamma \right)^{1/2}$$

(6.17)

is the **unit cell volume**. In crystallographic textbooks, the dual basis vectors are often written as $\mathbf{a}^*$, $\mathbf{b}^*$ and $\mathbf{c}^*$.

Problem 6.2: show that the angle between the dual basis vectors $\mathbf{a}^*$ and $\mathbf{b}^*$ for a hexagonal lattice ($a = b \neq c$, $\alpha = \beta = 90°$, $\gamma = 12°$) is $60°$.

Problem 6.3: show that the angle between the real-space basis vector $\mathbf{a}$ and the dual basis vectors $\mathbf{a}^*$ for a hexagonal lattice ($a = b \neq c$, $\alpha = \beta = 90°$, $\gamma = 12°$) is $30°$.

6.2.4 A dummies' guide to calculating reciprocal lattice parameters

The lengths and angles between dual basis vectors, $a^*, b^*, c^*, \alpha^*, \beta^*, \gamma^*$, are known as **reciprocal lattice parameters**. Here is a simple way to calculate them from the direct lattice parameters:

- Calculate the **direct metric tensor** from Eqn. (6.5).
- Invert it and multiply by $(2\pi)^2$ to yield $\tilde{G}$. This is the **reciprocal metric tensor**.
- By analogy with Eqn. (6.5)

$$\tilde{G} = \begin{bmatrix} a^{*2} & a^*b^*\cos\gamma^* & a^*c^*\cos\beta^* \\ a^*b^*\cos\gamma^* & b^{*2} & b^*c^*\cos\alpha^* \\ a^*c^*\cos\beta^* & b^*c^*\cos\alpha^* & c^{*2} \end{bmatrix}. \quad (6.18)$$

- Equation (6.18) can be easily solved for $a^*, b^*, c^*, \alpha^*, \beta^*, \gamma^*$.
- The **reciprocal cell volume**, v^*, is

$$v^* = \frac{(2\pi)^3}{v}. \quad (6.19)$$

6.3 An auxiliary Cartesian system

In many cases, it is possible to identify a suitable Cartesian system that is related to the basis vectors via a simple transformation. This is useful in cases where vector relations are straightforward in Cartesian coordinates but not obvious in generalized coordinates (see below, for example, for the case of the cross-product). A suitable basis can be found by projection (Figure 6.1).

One can always choose the new a-axis, $\mathbf{a}_c$, *parallel* to $\mathbf{a}$. Also, one can notice that $\mathbf{c}_c$ is *perpendicular* to both $\mathbf{a}$ and $\mathbf{b}$, so it is the obvious choice of the new c-axis, $\mathbf{c}^*$. Simple (but tedious) trigonometry can be used to find the relation between the two bases (for an extensive derivation, see Giacovazzo *et al.*, 2002, p 68):

$$[\mathbf{a}_c\ \mathbf{b}_c\ \mathbf{c}_c] = [\mathbf{a}\ \mathbf{b}\ \mathbf{c}]\, \mathbf{P}_c \quad (6.20)$$

$$\mathbf{P}_c = \begin{bmatrix} a^{-1} & -a^{-1}\cot\gamma & a^*/2\pi \cos\beta^* \\ 0 & b^{-1}/\sin\gamma & b^*/2\pi \cos\alpha^* \\ 0 & 0 & c^*/2\pi \end{bmatrix}$$

where a, b and c are the lengths of the basis vectors of the oblique system and α, β and γ are the angles between them.

Equation (6.20) enables one to express all the relevant vector operations in terms of the original coordinates. In particular, the metric tensor in *any* coordinate system can be written as $\mathbf{G} = \mathbf{P}_c^T \mathbf{P}_c$.

Another interesting case that can be solved with the help of Eqn. (6.20) is that of the *vector* product. In Cartesian coordinates the vector product can be written using the Levi-Civita symbol:

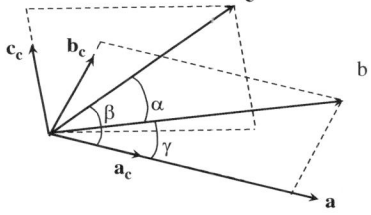

Fig. 6.1 Projective construction to obtain a Cartesian basis of unit length [$\mathbf{a}_c\ \mathbf{b}_c\ \mathbf{c}_c$] from a generic oblique basis [$\mathbf{a}\ \mathbf{b}\ \mathbf{c}$].

$$\varepsilon^i_{jk} = \begin{cases} +1 & \text{if } ijk \text{ is } (1,2,3),\ (2,3,1) \text{ or } (3,1,2), \\ -1 & \text{if } ijk \text{ is } (3,2,1),\ (1,3,2) \text{ or } (2,1,3), \\ 0 & \text{otherwise: } i=j \text{ or } j=k \text{ or } k=i \end{cases} \quad (6.21)$$

and

$$w^i = [\mathbf{v} \times \mathbf{v}]^i = \varepsilon^i_{jk} v^j u^k. \quad (6.22)$$

We now note that ε^i_{jk} is *covariant* in the indices j and k and *contravariant* in the index i. The expression for the vector product in a generic coordinate system is therefore

$$w'^i = \varepsilon'^i_{jk} v'^j u'^k \text{ where } \varepsilon'^i_{jk} = \varepsilon^l_{mn} Q^i_l P^m_j P^n_k. \quad (6.23)$$

Problem 6.4: *write the Levi-Civita symbol explicitly for hexagonal coordinates.*

A phase transition in two dimensions

7

In this section, we will exploit the concepts introduced in the previous sections to describe in detail a structural phase transition in two dimensions. For this purpose, we will consider a slightly modified version of the Escher drawing in Figure 3.16, which, as we have established, has symmetry *p3m*1 (No. 14). The new drawing is shown in Figure 7.1, and differs from the original one by the fact that all fishes and turtles pointing in the "SSE" direction have been lightened in color.

7.1 Low-symmetry group

The following observations can be made by inspecting the new pattern with the old symmetry superimposed (Figure 7.2):

1. All the three-fold axes are now lost, since they would relate creatures of different color.
2. The mirror planes running parallel to the SSE direction are retained. All the other mirror planes are lost.
3. The glide planes running parallel to the SSE direction are retained. All the other glide planes are lost.
4. The size of the unit cell is unchanged.

By combining these observation, one can readily determine the symmetry of the modified pattern to be *cm* (No. 5), which is *rectangular* with a (conventional) centered cell of *twice* the size of the original hexagonal cell. The primitive cells of the two systems coincide. It is noteworthy that this determination did not require introducing a coordinate system for either symmetry.

7.2 Wyckoff positions

We shall now examine, with the help of the ITC, how the symmetry of individual sites is modified by the phase transition. In the high-symmetry "phase", there are three distinct Wyckoff sites with symmetry $3m.$, labeled $1a$, $1b$ and $1c$, all with multiplicity 1. They cor-

68   *A phase transition in two dimensions*

Fig. 7.1 A modified version of the Escher drawing of fishes, birds and turtles. The original is in Figure 3.16.

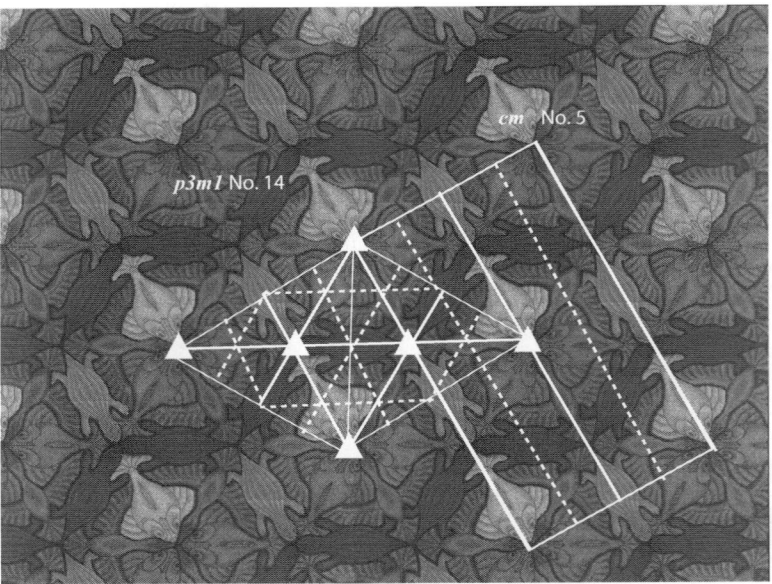

Fig. 7.2 The pattern in Figure 7.1 with the old *p3m*1 (No. 14) and the new *cm* (No. 5) symmetries superimposed.

respond to the heads of the fishes and birds (site $1a$ with our choice of origin), the tails of the turtles and birds (site $1b$) and the heads of the turtles/tails of fishes (site $1c$). All these sites are at the intersection of three mirror planes, of which one (parallel to the SSW direction) survives. We conclude that the symmetry of those sites below the phase transition will be $.m.$. The other special position in $p3m1$ is $3d$, with a local symmetry of $.m.$. Sites with this symmetry are located along the spines of the creatures. Moreover, one can see that each mirror plane defines the spines of all three creatures in the same succession, so there is only one type of $.m.$ site. In cm, there is only one special Wyckoff site ($2a$) with multiplicity 2 and symmetry $.m.$. However, we note that the size of the unit cell has been doubled, so the multiplicity in the primitive cell would be 1. These sites correspond to the spines of the creatures running in the SSW direction. The head/tail positions are no longer special, and have the same symmetry as the mirror they lie on. In the high-symmetry phase, general positions had multiplicity 6 (symbol $6e$). As an example, we can see that there are six identical front feet of the turtles arranged around their heads. In cm, the general position has multiplicity 4 (symbol $4b$, i.e., primitive multiplicity 2). This is because the darker turtles are no longer left–right-symmetric so that identical feet only come in pairs across the mirrors.

7.3 Basis transformation

Figure 7.3 shows a set of basis vectors for the high-symmetry ($p3m1$) and low-symmetry (cm) groups, both chosen according to standard crystallographic conventions. The a-axes of the two cells have been chosen to coincide, whereas the b-axis in the rectangular cell is the shortest translation orthogonal to the a-axis. We have therefore

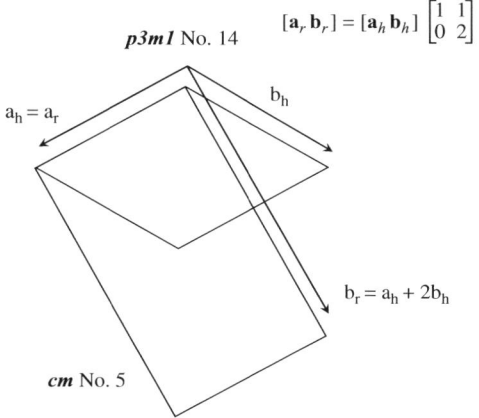

Fig. 7.3 Basis vectors for the $p3m1$ and cm wallpaper groups, set as in Figure 7.2.

$$\mathbf{a}_r = \mathbf{a}_h \tag{7.1}$$

$$|\mathbf{a}_r| = |\mathbf{a}_h| = a$$

$$\mathbf{b}_r = \mathbf{a}_h + 2\mathbf{b}_h$$

$$|\mathbf{a}_r| = a\sqrt{3}.$$

The covariant transformation is therefore:

$$[\mathbf{a}_r \ \mathbf{b}_r] = [\mathbf{a}_h \ \mathbf{b}_h]\begin{bmatrix} 1 & 1 \\ 0 & 2 \end{bmatrix} = [\mathbf{a}_h \ \mathbf{b}_h]\mathbf{P} \tag{7.2}$$

and the corresponding contravariant transformation is

$$\mathbf{Q} = \mathbf{P}^{-1} = \begin{bmatrix} 1 & -\frac{1}{2} \\ 0 & \frac{1}{2} \end{bmatrix}. \tag{7.3}$$

We can use this, for example, to determine the rectangular coordinates for the three-fold-axes positions in the original pattern, $1a = 0, 0$, $1b = \frac{1}{3}, \frac{2}{3}$ and $1c = \frac{2}{3}, \frac{1}{3}$. We find:

$$\begin{bmatrix} 1 & -\frac{1}{2} \\ 0 & \frac{1}{2} \end{bmatrix}\begin{bmatrix} 0 \\ 0 \end{bmatrix} = \begin{bmatrix} 0 \\ 0 \end{bmatrix} \tag{7.4}$$

$$\begin{bmatrix} 1 & -\frac{1}{2} \\ 0 & \frac{1}{2} \end{bmatrix}\begin{bmatrix} \frac{1}{3} \\ \frac{2}{3} \end{bmatrix} = \begin{bmatrix} 0 \\ \frac{1}{3} \end{bmatrix}$$

$$\begin{bmatrix} 1 & -\frac{1}{2} \\ 0 & \frac{1}{2} \end{bmatrix}\begin{bmatrix} \frac{2}{3} \\ \frac{1}{3} \end{bmatrix} = \begin{bmatrix} \frac{1}{2} \\ \frac{1}{6} \end{bmatrix}.$$

As we can see from the ITC, all three sets of rectangular coordinates correspond to Wyckoff positions $2a$.

7.4 Metric tensor

The wallpaper group cm has an orthogonal coordinate system, and it is therefore easy to determine its metric tensor:

$$\mathbf{G}_r = a^2 \begin{bmatrix} 1 & 0 \\ 0 & 3 \end{bmatrix}. \tag{7.5}$$

We can exploit the covariant properties of $\mathbf{G}$ to determine the corresponding metric tensor in hexagonal coordinates:

$$(\mathbf{G}_r)_{ij} = (\mathbf{G}_h)_{kl} P^k_i P^l_j = \mathbf{P}^T \mathbf{G}_h \mathbf{P} \tag{7.6}$$

whence

$$\mathbf{G}_h = \mathbf{Q}^T \mathbf{G}_r \mathbf{Q} = a^2 \begin{bmatrix} 1 & -\frac{1}{2} \\ -\frac{1}{2} & 1 \end{bmatrix}. \tag{7.7}$$

Let us verify the correctness of Eqn. (7.7). The distance between Wyckoff positions $1c$ and $1a$ is the length of the vector $[\frac{2}{3} \ \frac{1}{3}]$, the square of which is

$$d^2 = \begin{bmatrix} \frac{2}{3} & \frac{1}{3} \end{bmatrix} a^2 \begin{bmatrix} 1 & -\frac{1}{2} \\ -\frac{1}{2} & 1 \end{bmatrix} \begin{bmatrix} \frac{2}{3} \\ \frac{1}{3} \end{bmatrix} = \frac{1}{3} a^2. \tag{7.8}$$

In rectangular coordinates, we can do the calculation by hand:

$$d^2 = \left(\frac{1}{2}\right)^2 a^2 + \left(\frac{1}{6}\right)^2 3a^2 = \frac{1}{3} a^2. \tag{7.9}$$

SYMMETRY IN THREE DIMENSIONS

Part III

In this part of the book, we will employ the concepts we have learned and practiced for the simpler groups in 1D (frieze) and 2D (wallpaper) to tackle the more complex problem of describing the symmetry of "real" crystals in three dimensions. The 32 3D point groups are still relatively few and are simple enough to be mastered completely without the help of the ITC. However, a quick glance at the ITC diagrams for the more complex of the 230 space groups may lead the reader to despair that these can be understood, let alone derived without the help of the ITC. After reading these chapters, however, the reader may be pleasantly surprised by the ease with which many of the "simple" 1D and 2D concepts can be directly translated to 3D, enabling one to construct the diagrams and derive the symmetry properties of the *majority* of space groups from their Hermann–Mauguin symbol. Besides the obvious satisfaction of being able to impress a colleague at a conference by drawing a space-group diagram on a napkin without the help of the ITC, practical knowledge and familiarity with the space groups is invaluable in the study of phase transitions: one can often quickly establish, for example, how a structure can distort upon magnetic ordering simply by drawing the space-group diagram and "eliminating" the operators that are not compatible with the newly established spin structure. Clearly, these skills are very far removed from the kind of "automated" crystallography that modern diffractometers and software too often get us used to, but are still extremely useful in many branches of structural science.

Point groups in 3D

8

Having discussed in great detail the 2D case, we now move to analyze the 3D point groups, which represent the rotations of a 3D object around a fixed point that are mutually compatible with a system of translations. We will determine all possible 3D point groups in two steps: we will first derive all the 3D point groups having a 2D point group as a projection. In this case, each 3D symmetry operator is associated with a 2D symmetry operator acting on the *projection* of the solid of which the symmetry is sought onto a plane through the fixed point. We will then consider the remaining 3D point groups, which, as we shall see, correspond to the five cubic groups.

8.1 Generalized rotations in 3D

As already mentioned, all the crystallographically allowed generalized rotations in 3D can be obtained from the corresponding 2D operators by composition with the two operators E (the identity) and I (the inversion), both of them commuting with all operators. We have also seen that the composition of a two-fold axis with inversion is the mirror plane orthogonal to the axis. The inversion itself is represented graphically by a small circle (○), which can be combined with other symbols, if required (see, for instance, the roto-inversion $\bar{3}$ below). The remaining "roto-inversion" operators are $\bar{3}$ (▲), $\bar{4}$ (◊) and $\bar{6}$ (▲), and their action is summarized in Figure 8.1. The symbols are chosen to emphasize the existence of another operator inside the "belly" of each new operator. Note that $\bar{3}^3 = I$, and $\bar{3}^4 = 3$, i.e., symmetries containing $\bar{3}$ also contain the inversion and the three-fold rotation. Conversely, $\bar{4}$ and $\bar{6}$ do not automatically contain the inversion. In addition, symmetries containing both $\bar{4}$ (or $\bar{6}$) and I also contain 4 (or 6). Clearly, more orientations of all the allowed axes are possible than in the 2D case.

In the ITC, 3D point groups are represented by means of a *projection* onto a plane, which is chosen according to conventions. In all but two cases (see below), the axis of highest order is chosen to be *perpendicular* to this projection plane. In what follows, we will refer to the direction of the highest-order axis as the z-direction. Axes orthogonal to this axis are placed at the periphery of the projection circle (representing the

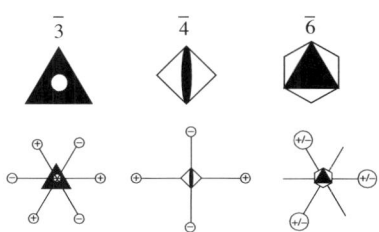

Fig. 8.1 Action of the $\bar{3}, \bar{4}$ and $\bar{6}$ operators and their powers. The set of equivalent points forms a trigonal antiprism, a tetragonally distorted tetrahedron and a trigonal prism, respectively. Points marked with "+" and "−" are above or below the projection plane, respectively. Positions marked with "+/−" correspond to pairs of equivalent points above and below the plane.

stereographic projection of a sphere), and the same symbols are used in all cases. If a mirror plane *perpendicular* to the vertical direction is present (known as a *horizontal* mirror plane, see below), this is indicated by drawing the projection circle with a **bold** line.

8.2 3D point groups with a 2D point-group projection

We can readily obtain all 3D point groups that have a 2D point group as a projection. By this, we mean that the transformation enacted by each 3D symmetry operator upon the *projection* of the object on a plane orthogonal to the highest-order axis is an allowed 2D symmetry operator. Clearly, these operators form a group, which must necessarily be one of the ten 2D point groups we already know. Hereafter, we will refer to these 3D point groups as "projective". It can be seen immediately that in order for a 3D group to be "projective" in the sense we just mentioned, the graph of the highest-order axis must be left invariant by all operators. In fact, all the points along this axis are projected on the invariant point of the 2D projection. Projection axes can therefore be of any order, but the only other allowed operators are two-fold proper and improper rotations *orthogonal* to the highest-order axis. Naturally, the original ten 2D point groups are part of the set of the "projective" 3D groups, and can be considered as the symmetry of 3D objects in which each "slice" along the z-direction transforms independently into itself. More generally, the only difference between the way the 3D operator and its associated 2D operator act must be the effect on the "z"-coordinate, the one that is projected out. This coordinate must be either left invariant or reversed into $-z$, since the z-coordinate of the fixed point must be unchanged. In other words, the two operators are either the same or they differ by the application of a mirror plane orthogonal to the highest-order axis, as in $m_z \circ g$. In order to obtain this set of 3D groups, we will therefore be allowed to replace each 2D operator g with itself, with $m_z \circ g$ or with both operators independently. This procedure is illustrated in Figure 8.2. Note that $m_z \circ 2_z = I$, $m_z \circ m_x = 2_y$ (likewise for other in-plane mirrors), $m_z \circ 3_z = \bar{6}$, $m_z \circ 4_z = \bar{4}$ and $m_z \circ 6_z = \bar{3}$. On this basis, we can construct a grid of "candidate" point groups and obtain the full list by simply excluding duplicate entries (see below).

8.2.1 Notation for "projective" 3D point groups

The notation we will employ is largely the same as that for 2D point groups. The only significant difference is that the presence of the mirror plane m_z *orthogonal* to a symmetry axis is denoted with the symbol "$/m$" right after the axis itself. There is also a new "rule of priority", stating that "un-barred" axes take precedence over "barred" ones, except for m, which takes precedence over 2. There are two forms of

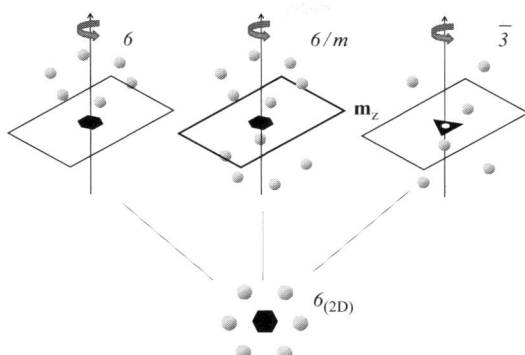

Fig. 8.2 Schematic representation of the method employed to generate, from 2D point groups, 3D groups that have that 2D group as a projection.

the Hermann–Mauguin symbols for point-groups: the "short form", in which only the highest-order axis is indicated (as in $4/mmm$), and the "long form", in which all axes are indicated explicitly (as in $\frac{4}{m}\frac{2}{m}\frac{2}{m}$).

8.2.1.1 The Schoenflies notation

The Schoenflies notation for 3D point groups (Schoenflies, 1891) is still widely used, and its knowledge is indispensable to understand older literature. In this notation, rotation and roto-inversion axes of order n are indicated with the symbols C_n and S_n, respectively ("C" for cyclic, "S" from the German word for mirror, "spiegel"). The symbol D_n (from "dihedron") indicate a cyclic axis with n two-fold axes orthogonal to it. The subscripts h (horizontal) and v (vertical) stand for mirror planes orthogonal and parallel, respectively, to the highest-order axis, whereas the subscript i stands for the inversion. Special symbols are used for point group $\bar{1}$ (consisting of the identity and the inversion, symbol C_i instead of C_{1i}) and point group m (consisting of the identity and a mirror plane, symbol C_s instead of S_2). Sometimes, more than one notation can be found for the same point group; the convention is to use C and D whenever possible in preference to S (which is in fact only used for point group S_4).

Problem 8.1: write the Schoenflies notations for the 27 "projective" point groups. In some cases, more than one notation can be found (e.g., S_3 and C_{3i}).

8.2.1.2 Axes conventions

Axes conventions are significantly more intricate than in the case of 2D point groups, where the first symbol always referred to the rotation axis, chosen to coincide with the z-axis. For 3D groups, the same convention applies to tetragonal (four-fold axis), trigonal (three-fold axis) and hexagonal (six-fold axis) point groups: the highest-order axis is listed first and is oriented in the c-direction. For monoclinic (two-fold), the two-fold axis (proper or improper) can coincide with the b-direction (standard setting) or the c-direction (alternate setting). For orthorhombic groups, the symbol refers to directions x, y and z in this order.

8.2.2 Tabular derivation of "projective" 3D point groups

As anticipated, we will derive the "projective" 3D point groups by systematically replacing symbols in the 2D point groups with operators having the same projection. Some symbols are subsequently excluded either because they are equivalent to point groups with a different orientation or because the composition of the generators with each other generates symbols with higher precedence. For example, $\bar{4}mm$ does not exist as a symbol, because, as we know, the composition of the two mirrors (at 45° of each other) gives a four-fold axis, which takes precedence over the $\bar{4}$. Likewise, $4/m2m$ and $4/m22$ cannot exist, because the composition of the inversion ($m_z \circ 4^2 = I$) with the two-fold axes gives vertical mirrors, which take precedence over the axes. Note also that $\bar{6}^3 = m_z$, so m_z always exists as an independent operator in the presence of $\bar{6}$ (and is therefore not listed). The resulting 27 3D point groups are listed in Table 8.1.

Table 8.1 The 27 "projective" point groups in 3D. These groups can be obtained from the 2D groups by considering the latter as their projections. The leftmost column contains a list of the ten 2D point groups, and the other three columns the "derived" 3D groups obtained by replacing operators systematically as explained in the text. "$m_z \circ n$" means that the highest-order axis has been replaced by its composition with m_z, whereas "m_z, n" means that both operators exist independently in the new group. Note in particular the 3D group $\bar{3}m1$, obtained from $\bar{6}mm$ by replacing 6 with its composition with m_z ($\bar{3}$) and one of the m with a two-fold axis. Because the two symmetry directions are orthogonal, the resulting symbol is $\bar{3}2/m1$, or $\bar{3}m1$ for short.

2D point group	n	$m_z \circ n$	m_z, n
1	1	m_z	—
2	2	$\bar{1}$	$2/m$
m	$m_x (= m_z)$	—	—
2mm	2mm	~~mmm~~	$2/mmm$
	~~2m2~~	$m2m(=2mm)$	~~2/m2m~~
	222	$m22$	~~2/m22~~
4	4	$\bar{4}$	$4/m$
	4mm	~~$\bar{4}mm$~~	$4/mmm$
4mm	~~$4m2$~~	$\bar{4}m2$	~~$4/mm2$~~
	422	$\bar{4}22$	~~$4/m22$~~
3	3	$\bar{6}$	—
3m1	3m1	$\bar{6}m2$	—
	321	$\bar{6}21 (= \bar{6}m2)$	—
6	6	$\bar{3}$	$6/m$
	6mm	~~$\bar{3}mm$~~	$6/mmm$
6mm	~~$6m2$~~	$\bar{3}2/m = \bar{3}m1$	~~$6/mm2$~~
	622	$\bar{3}22$	~~$6/m22$~~

8.3 Other 3D point groups: the five cubic groups

At this point, we may legitimately ask which other 3D point groups exist, beside the "projective" ones we just derived. In order to answer this question, we should observe that in-plane operators of the "projective" groups are proper or improper two-fold rotations, all forming angles of 90° with the highest-order axis. The missing groups will therefore involve at least two axes of order higher than two, either at 90° with each other or set at different angles. Given two rotations of angles α and β, with axes set at an angle θ with each other, one can show that the composition of the two rotations is a new rotation through an angle, γ, related to the other angles by:

$$2\cos\gamma + 1 = \cos\alpha\cos\beta\left(1 + \cos^2\theta\right)$$
$$-2\sin\alpha\sin\beta\cos\theta + (\cos\alpha + \cos\beta)\sin^2\theta + \cos^2\theta. \quad (8.1)$$

One can test Eqn. (8.1) with axes of different order, and find appropriate angles θ between them so that γ is a crystallographically allowed axis. Further, one must test the powers of the rotations (e.g., 2α with β, α with 2β, etc.) and verify that they too give an allowed rotation. These are very strict conditions, and are satisfied only by two types of rotations:

1. Three-fold axes set at 70.53° ($\cos\theta = \frac{1}{3}$), as the diagonals of a cube. Composition of two such rotations in the *same* direction gives a two-fold axis through one of the cube faces. Composition in the *opposite* direction yields another three-fold axis. By subsequent composition and graph symmetry, one retrieves all four three-fold axes through diagonals of a cube and three two-fold axes through its faces.
2. Four-fold axes set at 90°, as through the faces of a cube. Composition in *any* direction gives a three-fold axis through the cube diagonals. By subsequent composition and graph symmetry, one retrieves all four three-fold axes and three four-fold axes, plus six two-fold axes through the cube edges.

From these two groups, constructed with proper rotations only, plus compositions with inversion, one can obtain the five cubic point groups. Their Hermann–Mauguin symbols are similar to those of the other groups, with the cube faces as primary symmetry directions, the cube diagonal as secondary and the cube edges as tertiary (Figure 8.3). The Schoenflies symbol is T (for tetrahedral) or O (for octahedral) depending on the absence or presence of proper four-fold rotations.

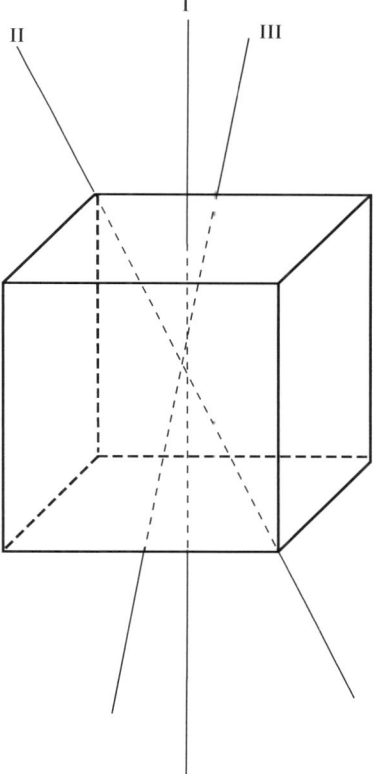

Fig. 8.3 The symmetry directions of a cube: primary (I) – three four-fold axes through the faces; secondary (II) – four three-fold axes through the corners; and tertiary (III) – six two-fold axes through the edges.

23 (Schoenflies notation T). This corresponds to the group described in item 1 above, and is the symmetry of a "chiral" tetrahedron (e.g., with faces marked with a three-fold "propeller").

$2/m\bar{3}$ ($m\bar{3}$ for short, Schoenflies notation T_h). The same generators as 32 plus inversion. It is the symmetry of a double tetrahedron yielding a centrosymmetric solid.

432 (Schoenflies notation O). This corresponds to the group described in item 2 above, and is the symmetry of a "chiral" cube, for example, with faces marked with a four-fold "propeller". Note that the two-fold axis along the tertiary direction is obtained by composition of the four-fold axis (say along the z-direction) with a 4^2 two-fold axis (say, along the x-direction).

$\bar{4}3m$ (Schoenflies notation T_d). This is the full symmetry of the tetrahedron. It is obtained from the previous group by replacing the four-fold axis with $\bar{4}$. By the previous argument, the tertiary two-fold axes are now replaced by mirrors.

$4/m\bar{3}2/m$ ($m\bar{3}m$ for short, Schoenflies notation O_h). This represents the full symmetry of a cube or octahedron.

Problem 8.2: find other examples of solids having the symmetry of the five cubic point groups.

8.4 3D point groups in the ITC

A full presentation of the 3D point groups is given in the ITC, Volume A (Hahn, 2002), pp. 770–790. The groups are arranged in *crystal systems* (first introduced by the German mineralogist Christian Samuel Weiss, 1780–1856 (Weiss, 1815, 1817)[1]), based on the symmetry of the lattices they support (see below). The following points summarize the few symbols and conventions used in these tables that have not been previously introduced.

- In all cases, the projection axis is the z-axis, regardless of the group setting (see below).
- In the left-handed projection, dots refer to the general position (or face poles) *above* the projection plane, whereas open circles are *below* the projection planes.
- *Monoclinic* groups (2, m and $2/m$) are presented in two different setting, with the "unique" axis (i.e., the direction of the proper or improper rotation) along either b or c.
- *Trigonal* groups (3, $\bar{3}$, 321, $3m1$ and $\bar{3}m1$) are each presented in three different settings. In addition to the two settings already present in the 2D case (i.e., with the hexagonal axes either parallel or perpendicular to the secondary symmetry direction), there is a

---

[1] In these papers, the monoclinic and triclinic systems were not described correctly. The correct crystal systems were established a few years later by Friedrich Mohs, 1822.

third setting, labeled "rhombohedral". Here, only three semi-axes are shown, meaning that the three basis vectors are coming *out* of the projection planes in the direction shown (see below for the definition of a rhombohedral lattice).

- *Rule of priority*. When more than one kind of symmetry operator occurs for a given symmetry direction, the priority rule states which operator takes precedence in the Hermann–Mauguin symbols. We state here the general rule of priority, which is valid for both point groups and space groups (clearly, roto-translation symbols are ignored for the former). The choice for reflections is made in order of descending priority:

$$m, e, a, b, c, n, d.$$

Rotation axes take priority over screw axes, which take priority over roto-inversions (with the exception of $\bar{3}$, which takes priority over 3).

The 14 3D Bravais lattices

9.1 Introduction

Since the reader of this book will most likely be familiar with the 14 Bravais lattices, the emphasis of this chapter will be on showing how these lattices are constructed and why no more (and no less) are needed. The procedure followed to derive the 14 Bravais lattices in 3D is closely related to that used in the 2D case. In fact, we can exploit the symmetry of the projections along symmetry directions, which, as we will see, must be one of the known 2D lattices.

In all but the trivial cases of the triclinic groups 1 and $\bar{1}$, where the three shortest independent translations are allowed to be at any angle with each other, there is at least one proper or improper axis of order 2 or higher. Following the procedure shown in Figure 3.4, one can always construct one translation *along* and two translations *perpendicular to* the axis direction, neither necessarily minimal, and in general not orthogonal to each other. Pairs of such translations define a 2D lattice, which must be one of the five lattices defined above. The various cases are obtained by considering all the possible directions of the minimal translation vectors.

9.2 Construction of the 14 Bravais lattices

9.2.1 Triclinic system

9.2.1.1 *Classes 1 and $\bar{1}$, holohedry $\bar{1}$, lattice P.*

There is no symmetry restriction on the basis vectors, which are therefore allowed to be at any angle with each other. The lattice symbol is P.

9.2.2 Monoclinic system

9.2.2.1 *Classes 2, m and 2/m, holohedry 2/m, lattices P and C*

The holohedry is easily derived by completing each of the classes with the inversion, which is necessarily a symmetry of the lattice. The lattice orthogonal to the two-fold axis (the "unique" axis) is oblique, whereas the other two are rectangular, and can be either p or c. This is shown in Figure 9.1

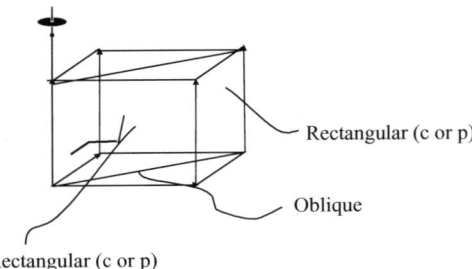

Fig. 9.1 Construction of 2D lattices as sublattices of a 3D monoclinic lattice. Lattices *containing* the two-fold axis are rectangular and can be p or c, whereas lattices *perpendicular to* the two-fold axis are oblique, and can only be p.

There are two possibilities:

- The shortest translation *not* orthogonal to the unique axis is also parallel to it. This means that all the rectangular lattices containing the two-fold axis are p. The translation along the two-fold axis is chosen as the **b** basis vector (or **c** in the alternate setting). In this case, one can choose as a primitive (and also conventional) cell a prism with parallelogram basis. The edges of the latter are chosen as **a** and **c** (or **a** and **c**) basis vectors. The lattice is *monoclinic primitive* (symbol P).
- The shortest translation *not* orthogonal to the unique axis is in an arbitrary direction. Twice its projection on the two-fold axis must however be a translation vector, and is chosen as the *conventional* **b** basis vector. Twice the projection on the plane perpendicular to the unique axis (the "basal plane") is also a translation vector; the standard setting identifies this direction with the **a**-axis. The **a** and **b** basis vectors define a c rectangular lattice containing the unique axis (the centering vector is the shortest translation *not* orthogonal to the unique axis). The **c**-axis is chosen as a vector in the basal plane that is linearly independent of **a**. With these conventions, the lattice is *monoclinic C-centered* (symbol C), because only the 2D lattice not containing the c-axis is C-centered. However, a plethora of other settings is possible, most of them duly reported in the ITC, making the monoclinic system the most intricate as far as conventions are concerned.

Problem 9.1: show that any monolinic lattice can be reduced to one of the two types C and P – see Figure 9.2.

9.2.3 Orthorhombic system

9.2.3.1 *Classes* 222, *mm2 and mmm, holohedry mmm, lattices P, C, F and I.*

The holohedry is easily derived again by completing each of the classes with the inversion. Once again, it is possible to find three translations

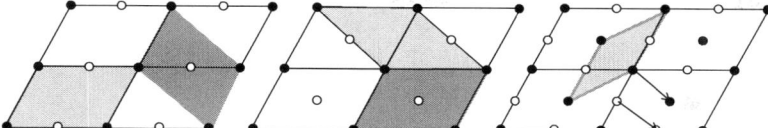

Fig. 9.2 Choices of monoclinic unit cells. The drawing shows that a C-centered cell can always be chosen regardless of which of the rectangular lattices containing the unique axis is centered. Filled circles are at $z = 0$; empty circles are at $z = \frac{1}{2}$. Left: the conventional C-centered monoclinic cell. Middle: a hypothetical I-centered cell (with a lattice point at $[\frac{1}{2}, \frac{1}{2}, \frac{1}{2}]$); the lattice is still C-centered with a different choice of unit cell. Right: a hypothetical F-centered cell (with lattice points at $[\frac{1}{2}, 0, \frac{1}{2}]$ and $[0, \frac{1}{2}, \frac{1}{2}]$); again, a C-centered choice of unit cell is shown.

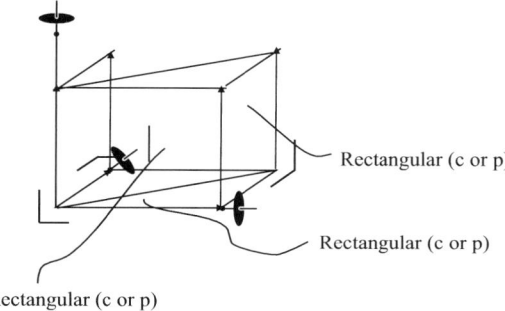

Fig. 9.3 Construction of 2D lattices as sublattices of a 3D orthorhombic lattice. All 2D lattices are rectangular, and can be p or c.

coinciding with the three rotation axes, and these are chosen as edges of the *conventional* unit cell. All the lattices orthogonal to these directions are rectangular, and so are lattices bisecting the axes, because they contain a two-fold axis (identical to m in 2D) – see Figure 9.3. It is useful to represent lattice points in fractional coordinates to check for compatibilities between 2D lattices. There are four possibilities:

- All of these 2D lattices are p. The primitive unit cell is a cuboid, and the lattice is "P-orthorhombic".
- One of the lattices orthogonal to a two-fold axis is c. The resulting 3D lattice is A, B or C depending on the orthogonal direction, clearly a matter of convention. The C setting is the standard one, but A also occurs as a standard setting in the $mm2$ class.
- All three lattices orthogonal to two-fold axes are c (it can be shown that if two of them are c, the third must also be c). The lattice is F, indicating that all faces are centered.
- One of the lattices bisecting two two-fold axes is c. By using fractional coordinates, it can be shown that none of the other lattices can be c. In fact, if $[\frac{1}{2} \frac{1}{2} \frac{1}{2}]$ is a lattice point, then $[\frac{1}{2} \frac{1}{2} 0]$ and permutations cannot be lattice points, otherwise (by difference) $[0\,0\,\frac{1}{2}]$ would be a lattice point, contradicting the fact that $[0\,0\,1]$ is the shortest trans-

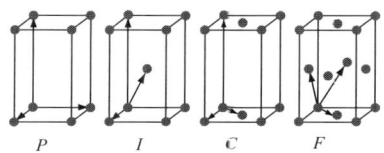

Fig. 9.4 The four orthorhombic lattices (see text).

lation along the z-axis. The lattice is body-centered orthorhombic (symbol I).

9.2.4 Tetragonal system

9.2.4.1 *Classes* 4, $\bar{4}$, 422, 4/m, 4mm, $\bar{4}m2$, 4/mmm, *holohedry* 4/mmm, *lattices P and I*.

The basal-plane lattice is square, and can only be p. The lattices orthogonal to the shortest in-plane translations are rectangular, but can only be p. In fact, if $[\frac{1}{2} 0 \frac{1}{2}]$ existed, so would $[0 \frac{1}{2} \frac{1}{2}]$ (by symmetry) and $[\frac{1}{2} -\frac{1}{2} 0]$ (by difference), contradicting the fact that $[1\,0\,0]$ is the shortest in-plane translation. This leaves only two possibilities:

- All of the 2D lattices are p. The primitive unit cell is a square-based cuboid, and the lattice is P-tetragonal.
- The rectangular lattices bisecting the in-plane directions are c. The lattice is body-centered tetragonal (symbol I).

Problem 9.2: show that a tetragonal F lattice can be reduced to I by an appropriate change of the unit cell.

9.2.5 Trigonal system

9.2.5.1 *Classes* 3, 3m1, 321, $\bar{3}m1$, *lattices P and R*.

This system is peculiar, in that each class can be supported by two lattices, P and R, with different holohedries.

- The P lattice is simply the 3D extension of the 2D hexagonal lattice by a translation along the z-axis, and has holohedry 6/mmm. Here, the unit cell is a hexagonal prism.
- The R lattice is generated from an arbitrary translation by applying the point group $\bar{3}m1$, the equivalent points forming a trigonal antiprism. Differences between translations with the same z-components are in-plane translations, forming a hexagonal lattice (Figure 9.5). Sums of three symmetry-equivalent translations with the same z-component are translations along the z-axis, so the z-component of any translation must be an integral multiple of $\frac{1}{3}$ of a lattice vector. The *primitive* unit cell is a *rhombohedron*, i.e., a cube "stretched" along one of the body diagonals. The larger (three times the volume) hexagonal cell, formed by the in-plane translations and by the shortest z-axis translation, is used as a "conventional" cell in the "hexagonal" setting. Both rhombohedral and hexagonal settings are used and are listed in the ITC.

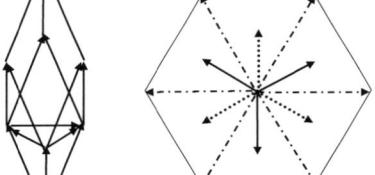

Fig. 9.5 Left: Composition of a generic translation with its symmetry-equivalents through the point group $\bar{3}m1$. The figure emphasizes the in-plane "difference" translations, forming a hexagonal lattice, and the edges of the rhombohedral unit cell. The height of the rhombohedron is the shortest translation along the z-axis, and is used as the **c** basis vector in the hexagonal setting. Right: projection of part of the same construction on the basal plane, emphasizing the "conventional" hexagonal cell and its relation with the primitive translations. Continuous arrows point out of the projection plane; dotted arrows point into the projection plane; dot-dash arrows are in the plane and form a hexagonal lattice.

9.2.6 Hexagonal system

9.2.6.1 *Classes 6, $\bar{6}$, 6/m, 622, $\bar{6}m2$, 6mm, 6/mmm, holohedry 6/mmm, lattice P.*

It is easy to see by subtraction of equivalent translations that the projections of a generic translation in the basal plane and perpendicular to it are also translations. The only allowed lattice is therefore P, and the unit cell is a hexagonal prism.

9.2.7 Cubic system

9.2.7.1 *Classes 23, $m\bar{3}$, 432, $\bar{4}3m$ and $4\bar{3}m$, holohedry $4\bar{3}m$, lattices P, I and F.*

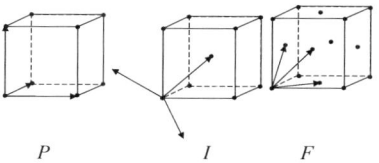

P   I   F

Fig. 9.6 Primitive and conventional cells for the three cubic lattices: primitive – P (left); body-centered – I (middle); and face-centered – F (right).

The three lattices are generated by the shortest translations and its equivalents, which, by restriction, must be along one of the three inequivalent symmetry directions. In all cases, the cubic cell is used as the conventional cell, and the edges of the primitive cell are obtained by rotating the generating translation around a three-fold axis (Figure 9.6).

- The shortest translation is along the *primary* direction (four-fold axis). The lattice is *primitive cubic* (P), and the primitive unit cell is a cube.
- The shortest translation is along the *secondary* direction (three-fold axis), and connects a corner with the center of the conventional cubic cell. The primitive cell is a rhombohedron with angles between edges $\alpha = 109.47°$ ($\cos \alpha = -\frac{1}{3}$). The lattice is I – "body-centered cubic".
- The shortest translation is along the *tertiary* direction (two-fold axis along the face diagonals). The primitive cell is a rhombohedron with angles between edges $\alpha = 60°$. The lattice is F – "face-centered cubic".

3D space-group symmetry

10.1 Roto-translations in 3D

We are now ready to introduce the remaining symmetry operators in 3D, not previously seen either as pure proper or improper rotations or as compositions of rotations with orthogonal translations. These remaining operators will therefore be compositions of proper rotations with translations *parallel* to them, and are known as *screw axes*. "Improper screw axes" do not exist as distinct roto-translation operators, because they have invariant points (see below — remember that roto-translation are characterized by the absence of invariant points). The following properties, relating to the composition of translations with proper and improper rotations in 3D, can be easily proven:

10.1.0.1 *Screw axes*

- For screw axes, the translational component commutes with the rotational component, since the former is invariant by rotation.
- For an axis of order n, the n-th power of a screw axis is a lattice translation. This is obvious if one shifts all the translational parts to the right, by exploiting the commutation properties. Therefore, the translation component must be $\frac{m}{n}t$, where m is an integer. We can limit ourselves to $m < n$, all the other operators being composition with lattice translations. Roto-translation axes are therefore indicated as n_m, as in 2_1, 6_3, etc.
- Screw axes can be chiral. This means that the set of operators sharing the same graph (the roto-translation and its powers) is distinct from its mirror image. Chiral space groups contain a chiral axis but do not contain reflections (mirrors and/or glides). They always come in *enantiomorphic* pairs (e.g., $P4_12_12$ and $P4_32_12$). If reflections are present, the space group will contain both types of chirality.

10.1.0.2 *Improper rotations*

We recall that improper rotations, such as mirror planes and roto-inversions, are compositions of the *inversion* operator with a rotation axis.

- The composition of a translation with the inversion, as in $t \circ I$, is an inversion center translated by $t/2$ with respect to the original one. This can be easily seen as follows. Let $[I]$ be the graph of I, i.e., its invariant point. Let us define the new point $[I] + t/2$ and apply the operator $t \circ I$. We obtain

$$t \circ I[[I] + t/2] = t[[I] - t/2] = [I] + t - t/2 = [I] + t/2. \quad (10.1)$$

- The composition of an improper rotation of *any* order with a translation parallel to it is the same improper rotation with the invariant center displaced by $t/2$. In fact let p be the invariant point of the improper rotation $I \circ n$. Noting that the translation $t/2$ is invariant with respect to n,

$$t \circ I \circ n[[p] + t/2] = t[[p] - t/2] = [p] + t/2. \quad (10.2)$$

Clearly the mirror plane is included in this case, since $\bar{2} = m$.
- Pure rotation of any order and non-chiral roto-translations (2_1, 4_2 and 6_3) can coexist with a center of symmetry on their graph. This is not the case for chiral roto-translations, for obvious reasons.

10.2 Glide planes in 3D

Glide planes in 3D are not essentially different from the 2D case, since they are compositions of a mirror plane with a translation in the same plane, and their square is a pure lattice translation. However, since mirror planes in 3D are truly two-dimensional (unlike the 2D mirror "lines"), the translation can be oriented at different angles with respect to the symmetry directions lying in the plane.

a, *b*, or *c* The glide translation is parallel to one of the in-plane basis vectors of the conventional cell, and half of their length.

n The "diagonal" glide translation is *half the sum* of the in-plane basis vectors (e.g., $\frac{1}{2}\mathbf{a} + \frac{1}{2}\mathbf{b}$).

e This symbol for "double" glide planes stands for the graph of two distinct glide planes, each with glide translations parallel to an in-plane basis vector and half of it in length. These operators only occur in centered cells, because the composition of the two operators gives a pure translation $\frac{1}{2}\mathbf{a} + \frac{1}{2}\mathbf{b}$, i.e., a centering translation.

The symbol "*e*" is new to the fifth edition of the ITC (Hahn, 2002). In previous editions, only one of the glides was indicated ("*a*" or "*b*" in the standard setting).

d For this so-called "diamond glide", the glide translation is *half of a centering translation*. Because the projections of the diamond glide translations along the symmetry directions are $\frac{1}{4}$ of the lattice translations, diamond glides are especially restricted, and occur in very few space groups. Naturally they only occur in centered cells, but, for example, they can never be orthogonal to mirror planes.

Problem 10.1: show that the composition of two orthogonal glide planes is either a two-fold axis or a two-fold screw, depending on the direction of the glide vectors. Where will the new axis be located with respect to the intersection of the two glide planes?

10.3 Graphical notation for 3D symmetry operators

A table with all the graphical symbols for the 3D symmetry operators is given in the ITC, Volume A (Hahn, 2002) on pages 7–10. The most important symbols that we have not encountered so far are collected in Figure 10.1. Another important "novelty" in the 3D case is the use of

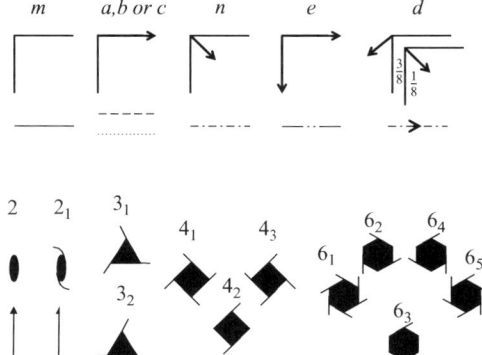

Fig. 10.1 Top: graphical symbols for mirror and glide planes. From left to right: *mirror planes m* – top and side view; *glide planes a, b and c* – top view and side view with glide translation in the plane of the sheet (dashed line) and orthogonal to it (dotted line); *diagonal glide n* – top and side view; *double glide plane e* – top and side view; *pair of diamond glides d* – top and side view. Bottom: roto-translation screw axes of all orders. Two-fold axes are also shown in projection.

fractions to indicate the vertical fractional coordinate of features such as inversion centers and horizontal reflection planes and axes (see, for example, the diamond glides in Figure 10.1).

10.4 Multiplication tables in 3D

We will not examine the full 3D multiplication table in detail, as we have done for the 2D case. Instead, we will provide a general recipe to deduce 3D multiplication "fragments" for the "projective" classes, which we have introduced previously. The basic idea is similar to the one we employed to construct the 3D point groups: for the projective classes, each 2D operator (including the roto-translations) can be thought of as the "projection" of one or more 3D symmetry operators onto an arbitrary horizontal plane. For example, a glide plane g orthogonal to the x-direction in 2D can be thought of as the projection of the glide planes a or n or of the roto-translation 2_1 parallel to the y-direction. All these operators differ only by their effect on the z-coordinate. One first set of 3D fragments can therefore be obtained by replacing each 2D symbol with 3D operators having the same projection. Other fragments can subsequently be generated by adding a horizontal mirror plane or glide and/or its composition with the highest-order axis, as in the case of the 3D point groups.

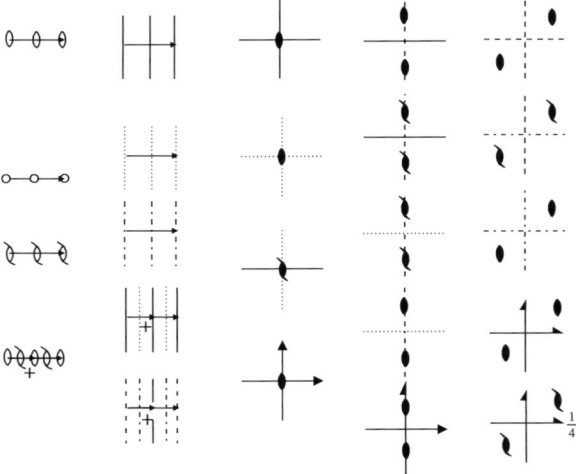

Fig. 10.2 Example of the construction of the 3D multiplication table for the classes 222 and $mm2$. Centering translations of the face perpendicular to the vertical axis are indicated with a "+". The two leftmost columns refer to composition of two-fold proper and improper rotations with translations. In the remaining three columns, all the fragments have the topmost fragment as 2D projection.

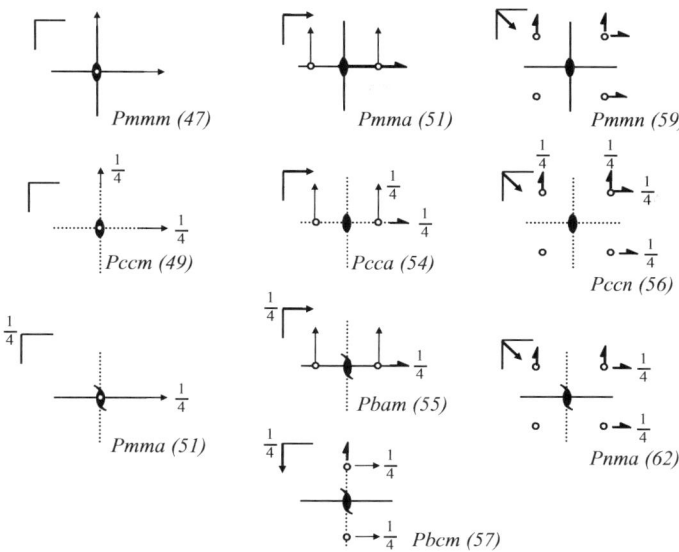

Fig. 10.3 Examples of the construction of part of the 3D multiplication table for the class *mmm*. The fragments in column 3 of Figure 10.2 have been combined with mirror planes (left), *a* or *b* glide planes (center) and *n* glide planes (right). An example of space groups where each fragment can be found is also listed. The origin has been shifted so that all the inversion centers are at $z = 0$.

Examples of this procedure, limited to the orthorhombic classes 222, *mm*2 and *mmm*, are shown in Figures 10.2–10.3.

Figure 10.4 shows more complex combinations of centering translations with the higher-order axes 3 and 4.

10.5 Construction of space groups in 3D

The complete enumeration of the 230 space groups or "groups of motion", as they were called at the time, was carried out independently and almost simultaneously by the Russian mineralogy and crystallographer E.S. Fedorov (1853–1928) (Fedorov, 1890, 1891), by the German mathematician Arthur Schoenflies (1853–1928) (Schoenflies, 1891) and by the "amateur" British scientist William Barlow (1845–1934) (Barlow, 1884). With some practice, it is possible to construct most of the 230 space-group graph diagrams, starting from the short form of the Hermann–Mauguin symbol. Most of the "projective" classes are quite simple to construct, and some of the cubic groups are also easily constructed, although for the most complex ones one is well advised to refer to the ITC. The construction procedure is

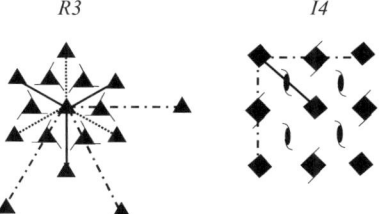

Fig. 10.4 Composition of higher-order axes with centering translations, as found in *R*3 and its supergroups (left) and *I*4 and its supergroups (right). Translations with components coming *out of* and *into* the projection plane are indicated with continuous lines and dotted lines, respectively. Dot-dash lines are in-plane translations.

of course particularly simple for the 73 so-called *symmorphic* space groups, where the point-group symmetry of at least one special position coincides with the class point group. The Hermann–Mauguin symbol of symmorphic space groups does not contain roto-translations, although the group itself usually does. For symmorphic groups, it is sufficient to draw the appropriate point-group diagram and "propagate" all the symbols by graph symmetry and composition with the translations.

For *non-symmorphic* groups, the point-group diagram must be replaced by an appropriate fragment of the 3D multiplication table, which, in turn, must then be propagated with the lattice translations. Rather than giving a recipe, we will present some examples of how this is done in practice.

10.5.1 Construction of $C2/c$ (No. 15)

The method is illustrated graphically in Figure 10.5. The unique axis is b, and the other two axes are set as shown in (1). We start by placing the two-fold axis at the origin, and propagating it by graph symmetry and composition with the translations defining the sides of the conventional cell, as in (2) (the b-direction is not shown). In (3), we have applied the c glide plane: the composition between this plane and the two-fold axes is a center of symmetry displaced by $1/4$ along c. In the final step, (4), we apply the C centering translation, i.e., the translation vector $[\frac{1}{2}\,\frac{1}{2}\,0]$. The composition of this translation with the two-fold axes yields

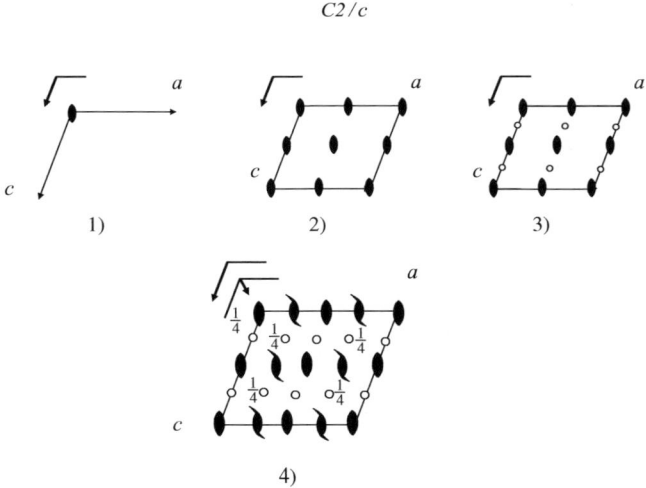

Fig. 10.5 Construction of the space group $C2/c$ (No. 15); see text for a detailed step-by-step description.

2_1-axes. The composition with the centers of symmetry yields other centers of symmetry displaced upwards (i.e., along b) by $\frac{1}{4}$. Finally, the composition with the c glide yields an n glide displaced upwards by $\frac{1}{4}$.

10.5.2 Construction of *Ama*2 (No. 40)

Our second example is for the orthorhombic space group *Ama*2 (No. 40). Like the previous case, this space group is not symmorphic. We start by intersecting the mirror plane m (perpendicular to a) with the glide plane a (perpendicular to b). This produces a 2D fragment that should be well known to us, and that we can complete with two-fold axes on the glide (step 2). These can then be propagated with the conventional translations. Finally (point 3) we can apply the centering translations. The component along the c-axis converts 2-axes in 2_1 and a glides into n.

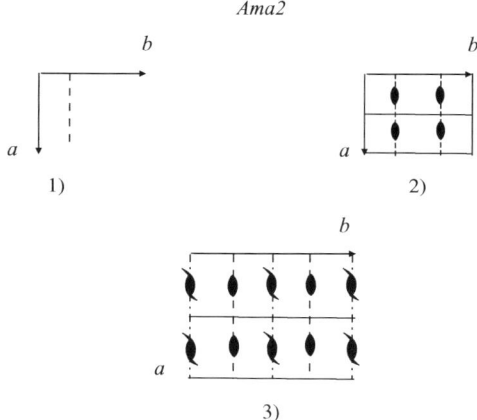

Fig. 10.6 Construction of the space group *Ama*2 (No. 40); see text for a detailed step-by-step description.

10.5.3 Construction of *P*4/*nmm* (No. 129)

This tetragonal example is considerably more complex. We start by constructing the subgroup *P*4*mm*, which is not only symmorphic but is also the exact replica of the 2D wallpaper group *p*4*mm*, with the exception of the lattice translations along the c-axis. This is done in Figure 10.7, steps (1) and (2). In step (3), we apply the horizontal glide n to the four-fold axes. We apply the mirror first, converting 4 into $\bar{4}$, then the translation, which places the $\bar{4}$-axes on top of the previous two-fold axes. We can also easily place the centers of symmetry. In the final

P4/nmm

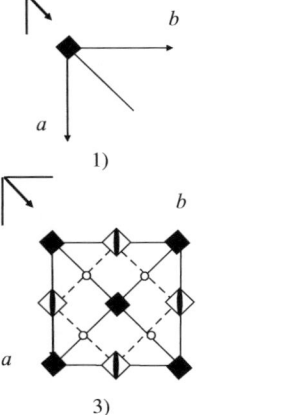
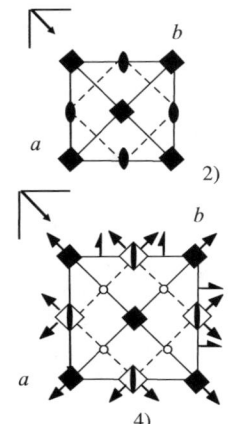

Fig. 10.7 Construction of the space group $P4/nmm$ (No. 129); see text for a detailed step-by-step description.

step (4), we compose the horizontal glide n with the vertical mirrors, obtaining two-fold axes of type 2 and 2_1.

10.5.4 Construction of $R3m$ (No. 160)

This example of a trigonal space group is quite straightforward if one starts from the construction in Figure 10.4, as in step (1). In step (2), we have replaced the translation vectors with the edges of the unit

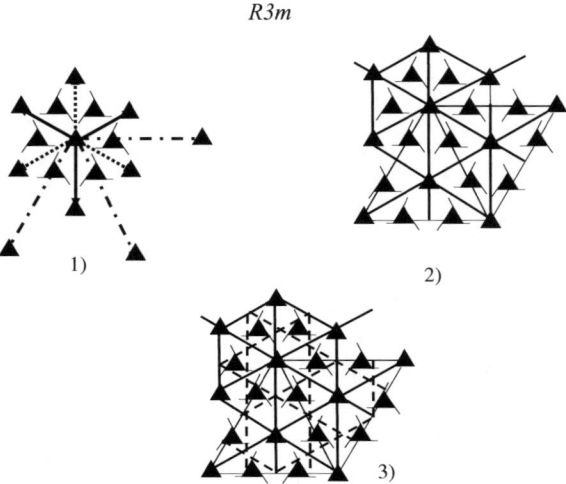

Fig. 10.8 Construction of the space group $R3m$ (No. 160); see text for a detailed step-by-step description.

cell. **Note**: the rhombohedral cell is seen from the top, so its edges meeting at the vertex point *down*. We have also added the mirror planes *m*, which have been simply propagated by graph symmetry. In step (3), the mirror planes have been composed with the *hexagonal* in-plane translations, yielding glide planes. It is noteworthy that a number of *additional* glides, marked *g* in the list of symmetry operators, exist with glide component $\frac{1}{3}$ or $\frac{2}{3}$ along the *c*-axis. Their graphs coincide with those of the mirror and in-plane glides, and do not have a separate graphical representation.

Problem 10.2: construct graphically the following space groups: C2/m, Cmca, Pnma and I4/mmm.

RECIPROCAL SPACE

Part IV

In this final part of the book, we will focus particularly on reciprocal space and its symmetry. This book is not about diffraction, and its purpose is to describe the general symmetry features of reciprocal space, not to derive scattering cross-sections for X-ray or neutron experiments. Nevertheless, we will introduce in a very general way the concept of the Fourier transform of lattice functions, and we will explain how these can be naturally assigned to the "nodes" of the reciprocal lattice, thereby producing a pattern that has no translational invariance but has nevertheless either the point-group symmetry of the crystal class or a higher symmetry (the Laue class). We will also describe in some detail the concept of *extinction* (or *reflection*) conditions, and explain how these conditions relate to the translational and roto-translational symmetry in real space. In the last chapter, we will discuss two topics that find little space in the ITC – they are briefly mentioned in Volume B (Shmueli, 2001) – but are essential to understand and handle the full symmetry properties of reciprocal space: the Wigner–Seitz construction and the extended Wigner–Seitz construction, also known as the Brillouin zone construction. The **Wigner–Seitz construction** enables one to construct a reciprocal-space unit cell (or a real-space unit cell, for that matter) that has the full point-group symmetry properties of the space as a whole. Starting from the Wigner–Seitz cell construction, one can use the **Brillouin zone construction** to "tile" the whole reciprocal space in such a way that each Brillouin zone (a subset of tiles) has the full symmetry properties of the reciprocal space. Moreover, each zone can be "reduced" to the first Wigner–Seitz unit cell (the one containing the origin of the reciprocal space) by translating it through a reciprocal lattice vector. These concepts are essential in describing dynamic phenomena in crystals – for example phonons, spin waves and the motion of electrons in metals and semiconductors.

Symmetry and reflection conditions in reciprocal space

11

11.1 Real and reciprocal lattice: recapitulation

11.1.0.1 *Real space and real lattice*
- **Real-space points** are described by means of an **origin** and **real-space position vectors**, as $p = o + \mathbf{v}$. The choice of the origin is *arbitrary*, and gives rise to sets of related position vectors.
- **Real-space position vectors** are generally described as linear combinations of the **real or direct basis vectors** (*covariant*, dimensions: length) with **dimensionless coefficients** (*contravariant coordinates*).
- **Real-space lattice (translation) vectors** are linear combinations of the **primitive basis vectors** with **integral components**. For certain lattices, they can also be expressed as linear combinations of the **conventional basis vectors** with **integral or simple fractional components**. Real lattice vectors with *fractional* components are known as **centering vectors**.

11.1.0.2 *Reciprocal space and reciprocal lattice (RL)*
- **Reciprocal-space position vectors** are described as linear combinations of the **reciprocal or dual basis vectors** (*contravariant*, dimensions: length^{-1}) with **dimensionless coefficients** (*covariant components*).
- **Reciprocal-space points** are obtained by adding the **reciprocal-space position vectors** to an **origin**, which, unlike the real-space origin, is **not arbitrary** (see below).
- **Reciprocal lattice vectors** are linear combinations of the **dual basis vectors** with **integral components**.

11.1.0.3 *Dot product*
- The dot product of *real* and *reciprocal space* vectors expressed in the usual coordinates is

$$\mathbf{q} \cdot \mathbf{v} = 2\pi q_i v^i. \tag{11.1}$$

- The dot product of *real* and *reciprocal **lattice*** vectors is:
  - **If a *primitive* basis is used to construct the dual basis,** 2π times an integer for *all* **q** and **v** in the real and reciprocal lattice, respectively. In fact, as we have just said, all the components are integral in this case.
  - **If a *conventional* basis is used to construct the dual basis,** 2π times an integer **or** a simple fraction of 2π. In fact, as we have just said, *the components of the **centering** vectors are fractional*.
- Therefore, if a **conventional** real-space basis is used to construct the dual basis, **only certain reciprocal-lattice vectors will yield a $2\pi n$ dot product with *all* real-lattice vectors**. It is quite easy to show (left as an exercise) that those reciprocal-lattice vectors **are exactly those generated by the corresponding primitive basis**.

A *conventional* basis generates *more* RL vectors than a corresponding *primitive* basis. As we shall see, the "*extra*" points are not associated with any scattering intensity – we will say that they are *extinct by centering*.

11.2 Reciprocal lattice – an alternative definition

In Section 6.2 (summarized in the previous section), we have briefly introduced the notion of reciprocal lattice as the set of *points* generated from vectors constructed by multiplying the dual basis vectors $\mathbf{b}_1$, $\mathbf{b}_2$, and $\mathbf{b}_3$ (or $\mathbf{a}^*$, $\mathbf{b}^*$ and $\mathbf{c}^*$ in the usual crystallographic notation) by integers, as in $h\mathbf{a}^* + k\mathbf{b}^* + l\mathbf{c}^*$. In order to carry out this construction we nominally need an origin that, for the sake of simplicity, we may take as being the same as that of the real-space lattice. We note, however, that the reciprocal lattice *vectors* (in short RLVs) do not depend on the choice of the real-space origin, but on the basis only (at least based on our present definition; below we will show that the choice of basis is largely irrelevant). A better way to look at the problem is to consider the RLV to be *reciprocal* to the real-space *translation vectors*. This is an implicit recognition that reciprocal "space" is a simple vector space, and does not require the affine space structure. In the remainder, we will say liberally that a vector belongs to the reciprocal lattice, meaning that is a RLV.

Taken at face value, our definition of reciprocal lattice depends on the choice of the covariant (real-space) basis vectors. However, it can be shown that the set of RLVs generated by all *primitive* basis vectors is the same. We introduce here an alternative definition that makes this fact transparent.

A vector belongs to the primitive *reciprocal lattice if and only if its dot product with all symmetry translation vectors is an integral multiple of* 2π.

It is easy to see that the condition stated above is necessary, provided that the generating basis is *primitive*. In fact, the dot product of vectors q and v, expressed in covariant and contravariant coordinates, respectively, is

$$\mathbf{q} \cdot \mathbf{v} = 2\pi q_i v^i. \qquad (11.2)$$

If the real-space basis is *primitive*, all the symmetry translations have integral coefficients, so clearly the dot products must all be integral multiples of 2π provided that the q_i are also integers. Conversely, the coefficients q_i can be obtained as

$$q_i = \frac{1}{2\pi} \mathbf{q} \cdot \mathbf{a}_i. \qquad (11.3)$$

If the dot product of $\mathbf{q}$ with any translation vector is an integral multiple of 2π, so must be the dot product with the basis vectors. It follows that all q_i are necessarily integers.

The new definition makes it clear that the RLVs generated from a primitive basis (we will call them the *primitive* RLVs) do not depend on the choice of basis, since the dot product is an invariant quantity. It is interesting to consider what happens when we use a *conventional* real-space basis to define the reciprocal basis. The reciprocal-space vectors hereby defined are a *superset* of the primitive RLVs – this can be clearly seen using a demonstration identical to the second part of the above demonstration. However, some *conventional* RLVs do not have dot products equal to integral multiples of 2π with *all* symmetry translations. In fact, centering translations have *fractional* components on the conventional basis. These additional vectors, which are part of the *conventional* but not of the *primitive* RLVs, are said to be *extinct by centering*. The reason of this name will become clear shortly.

11.3 Centering extinctions

As anticipated in the previous section, reciprocal space vectors generated by a conventional basis are said to be extinct by centering if their dot product with the centering vectors is not an integral multiple of 2π. These vectors are therefore not part of the *primitive* reciprocal lattice. Based on the known form of the centering vectors for the various lattices, we can easily find the form of these vectors. Because the conditions are expressed in fractional coordinates, the extinction conditions are the same for the same type of centering, regardless of the symmetry. The centering extinctions are summarized in Table 11.1.

A, B or C-centered lattices The centering vectors in conventional coordinates are $[0\,\frac{1}{2}\,\frac{1}{2}]$, $[\frac{1}{2}\,0\,\frac{1}{2}]$ and $[\frac{1}{2}\,\frac{1}{2}\,0]$, respectively, for the three types of lattice, modulo integral numbers. The dot product of a conventional RLV $\mathbf{q}$ with covariant coordinates h, k and l

Table 11.1 Centering extinctions and scattering conditions for the centered lattices. The "Extinction" columns lists the Miller indices of reflections that are **extinct by centering**, i.e., are "extra" RLVs generated as a result of using a conventional basis instead of a primitive one. The complementary "Scattering" column corresponds to the listing in the International Tables, Volume A (Hahn, 2002), and lists the Miller indices of "allowed" reflections. "n" is any integer (positive or negative).

Lattice type	Extinction	Scattering
P	none	all
A	$k + l = 2n + 1$	$k + l = 2n$
B	$h + l = 2n + 1$	$h + l = 2n$
C	$h + k = 2n + 1$	$h + k = 2n$
F	$k + l = 2n + 1$ or $h + l = 2n + 1$ or $h + k = 2n + 1$	$k + l = 2n$ and $h + l = 2n$ and $h + k = 2n$
I	$h + k + l = 2n + 1$	$h + k + l = 2n$
R	$-h + k + l = 3n + 1$ or $-h + k + l = 3n + 2$	$-h + k + l = 3n$

with these centering vectors is therefore $2\pi\frac{1}{2}(k + l)$, $2\pi\frac{1}{2}(h + l)$ and $2\pi\frac{1}{2}(h + k)$. We conclude that conventional RLVs with $k + l \neq 2n$, etc., where n is any positive or negative integer, are extinct by centering.

F-centered lattices Here, all the faces are centered, so any condition $k + l \neq 2n$, $h + l \neq 2n$ and $h + k \neq 2n$ will lead to extinction. Conversely, $k + l = 2n$, $h + l = 2n$ and $h + k = 2n$ must be simultaneously true for a conventional RLV to be part of the primitive reciprocal lattice.

I-centered lattices Here, the centering translations are of the type $[\frac{1}{2}\,\frac{1}{2}\,\frac{1}{2}]$, and their dot product with a conventional RLV $\mathbf{q}$ is $2\pi\frac{1}{2}(h + k + l)$. Conventional RLVs with $h + k + l \neq 2n$ are therefore extinct by centering.

R-centered lattice in hexagonal coordinates We recall that the rhombohedral cell is *primitive*, so it does not give rise to extinctions by centering. In the hexagonal setting, there are two centering translations – $[\frac{2}{3}\,\frac{1}{3}\,\frac{1}{3}]$ and $[\frac{1}{3}\,\frac{2}{3}\,\frac{2}{3}]$. The dot products $\mathbf{q} \cdot \mathbf{v}$ are $2\pi\frac{1}{3}(2h + k + l)$ and $2\pi\frac{1}{3}(h + 2k + 2l)$, and both must be integral multiples of 2π for $\mathbf{q}$ to belong to the primitive RL. This is equivalent to $2h + k + l = 3n$, $h + 2k + 2l = 3n$. We now note that if the first condition is satisfied, so is the second. In fact, $h + 2k + 2l = 3h + 3k + 3l - (2h + k + l)$, and both terms on the right side are three times an integer if $2h + k + l = 3n$. The only condition for belonging to the primitive reciprocal lattice is therefore $2h + k + l = 3n$, or also $-h + k + l = 3n$, since $3h$ is three times an integer.

11.4 Holohedry of the reciprocal lattice

With the new definition of RL, it is straightforward to prove that the RL has the same point-group symmetry (holohedry) as the real lattice. Let us assume that a given rotation operator **R** belongs to the holohedry of the real-space lattice. This is equivalent to saying that for each symmetry translation vector **t**, **Rt** is also a symmetry translation. Moreover, since each rotation possesses an inverse that is also part of the holohedry, *each* translation vector **t** can be expressed as **t** = **Rt**′, where obviously **t**′ = **R**$^{-1}$**t**. It follows immediately that if **q** is a primitive RLV, so is **Rq**. This arises from the fact that every rotation preserves the dot product. We can show this readily by using Cartesian coordinates, and since the dot product is invariant this statement must be true in any coordinate. In Cartesian coordinates,

$$\mathbf{Rq} \cdot \mathbf{t} = [Rq]^T [Rt'] = [q]^T R^T R[t'] = [q]^T R^{-1} R[t'] = \mathbf{q} \cdot \mathbf{t'}. \quad (11.4)$$

We can generalize Eqn. (11.4) to other coordinate systems. In particular, the invariance of the dot product by rotation of both **q** and **t** enables us to derive the rotation formula in reciprocal space:

$$(\mathbf{Rt})^i = R^i_j t^j \quad (11.5)$$

$$(\mathbf{Rq})_i = q_j (R^{-1})^j_i$$

$$\mathbf{Rq} \cdot \mathbf{Rt} = q_j (R^{-1})^j_i R^i_k t^k = q_j t^j = \mathbf{q} \cdot \mathbf{t}.$$

One consequence of the above statements is that the RL must necessarily be one of the 14 Bravais lattices. However, the fact that the real and reciprocal lattices have the same holohedry does not mean that the Bravais lattice type is the same. In addition, the conventional unit cell of the reciprocal lattice is larger than the one having the dual basis vectors as edges. It is left as an exercise to derive the RL types corresponding to the different real lattice types (for a summary, see Table 11.2).

Problem 11.1: show that for each Bravais lattice, the corresponding RL lattice is as follows:

106 Symmetry and reflection conditions in reciprocal space

Table 11.2 Reciprocal-lattice Bravais lattice for any given real-space Bravais lattice (BL).

Crystal system	Real-space BL	Reciprocal-space BL
Triclinic	P	P
Monoclinic	C	C
Orthorhombic	P	P
	A or B or C	A or B or C
	I	F
	F	I
Tetragonal	P	P
	I	I
Trigonal	P	P
	R	R
Hexagonal	P	P
Cubic	P	P
	I	F
	F	I

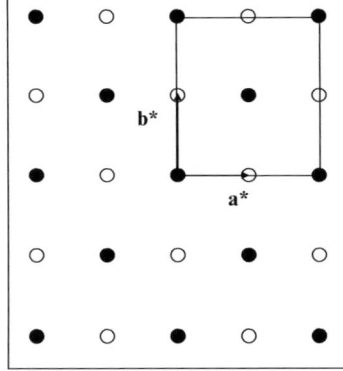

Fig. 11.1 Reciprocal lattice for the C-type orthorhombic and monoclinic lattices. Lattice points that are extinct due to centering ($h + k \neq 2n$) are shown as open circles. The reciprocal lattice is also C-centered, but its unit cell edges are twice the dual basis vectors.

A, B or C-centered lattices The RL is also A, B or C-centered, with the unit cell edges for the centered face given by $2\mathbf{a}^*$ and $2\mathbf{b}^*$ for the C-centered case (Figure 11.1).

I-centered orthorhombic or cubic lattices The RL is F-centered, with the unit-cell edges given by $2\mathbf{a}^*$, $2\mathbf{b}^*$ and $2\mathbf{c}^*$.

F-centered orthorhombic or cubic lattices The RL is I-centered, with the unit-cell edges given by $2\mathbf{a}^*$, $2\mathbf{b}^*$ and $2\mathbf{c}^*$.

I-centered tetragonal lattice The RL is I-centered, with the unit-cell edges given by $\mathbf{a}^* + \mathbf{b}^*$, $\mathbf{a}^* - \mathbf{b}^*$ and $2\mathbf{c}^*$ (Figure 11.2).

R-centered hexagonal lattice The RL is also R-centered, since the primitive dual cell is rhombohedral. The somewhat more complex relations between this cell and the dual hexagonal basis are

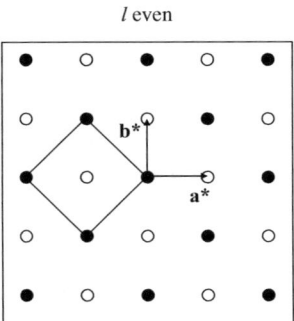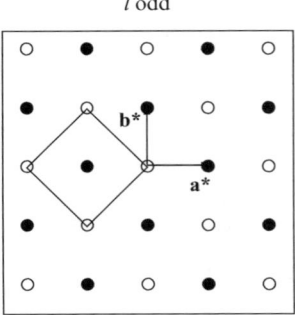

Fig. 11.2 Reciprocal lattice for the I-type tetragonal lattice. Lattice points that are extinct due to centering ($h + k + l \neq 2n$) are shown as open circles for l even (left) and odd (right). The reciprocal lattice is also I-centered, but its unit cell edges are given by $\mathbf{a}^* + \mathbf{b}^*$, $\mathbf{a}^* - \mathbf{b}^*$ and $2\mathbf{c}^*$.

illustrated in Figure 11.3. It is important to point out that the in-plane dual basis vectors for the hexagonal cell make an angle of 60° if the real-space basis is in the usual 120° setting. The edges of the R-centered conventional hexagonal cell of the RL are $2\mathbf{a}^* - \mathbf{b}^*$, $2\mathbf{b}^* - \mathbf{a}^*$ and $3\mathbf{c}^*$.

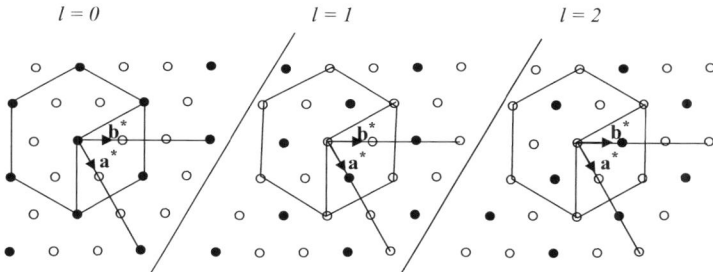

Fig. 11.3 Reciprocal lattice for the R-type trigonal lattice in hexagonal coordinates. Lattice points that are extinct due to centering ($-h + k + l \neq 3n$) are shown as open circles. The dual basis vectors in the hexagonal setting are shown as arrows (note the angle of 60°). The conventional hexagonal cell of the RL, with $2\mathbf{a}^* - \mathbf{b}^*$, $2\mathbf{b}^* - \mathbf{a}^*$ and $3\mathbf{c}^*$, is shown. The edges of a triple hexagonal cell with more complex centering rules is also shown, since it is sometimes referred to in textbooks.

11.5 Fourier transform of lattice functions: the "weighed" Reciprocal Lattice and its symmetry

In this section, we will consider a real or complex function $f(P)$ defined on the real-space points P. We assume that $f(P)$ has the symmetry properties defined by one of the 230 space groups. Once an origin O has been defined, $f(P)$ can be associated with an ordinary function $f(\mathbf{R})$ defined over the (three-dimensional) vector space $\mathbb{R}^3$, $\mathbf{R}$ being a position vector such that $P = P + \mathbf{R}$ (see Part II). Note that, with a different choice of the origin, the associated function $f(\mathbf{R})$ will be different. We will calculate the Fourier transform of this function, $F(\mathbf{q})$, and establish the effect of symmetry upon $F(\mathbf{q})$. We have:

$$F(\mathbf{q}) = \int d\mathbf{R}\, f(\mathbf{R}) e^{-i\mathbf{q}\cdot\mathbf{R}} \qquad (11.6)$$

where the integral extends to the whole space. Note that with the notation we adopted, some care is needed to ensure that the dimensions of $f(\mathbf{q})$ and $F(\mathbf{q})$ are consistent. $f(\mathbf{q})$ is typically chosen to be a *number density* (e.g., number of particles per unit volume), so that $F(\mathbf{q})$ is dimensionless.

11.5.1 Fourier transform expressed in crystallographic coordinates

In Eqn. (11.6), the Fourier transform is expressed in a concise and general form, which does not take into account the specific choice of coordinates. However, its expression in crystallographic coordinates is particularly simple: since the volume element in real space is $dxdydz(\mathbf{a} \times \mathbf{b}) \cdot \mathbf{c} = v_0 dxdydz$, v_0 being the volume of the unit cell, the Fourier transform becomes

$$F(\mathbf{q}) = v_0 \int\!\!\!\int\!\!\!\int_{-\infty}^{\infty} dxdydz f(x, y, z) e^{-2\pi i(xh+yk+zl)}. \qquad (11.7)$$

Hereafter, we will employ the abstract notation of Eqn. (11.6), knowing that the transformation to specific coordinates can always be made as in Eqn. (11.7).

11.5.2 Fourier transform and lattice periodicity

We now exploit the *lattice* periodicity of the function $f(\mathbf{R})$, which holds regardless of the choice of origin, and we can express this by writing $\mathbf{R} = \mathbf{R}_0 + \mathbf{r}$ and

$$f(\mathbf{R}_0 + \mathbf{r}) = f(\mathbf{r}). \qquad (11.8)$$

The $\mathbf{R}_0$ are the symmetry translation vectors. We can also decompose the integral in Eqn. (11.6) into integrals over the unit cells:

$$F(\mathbf{q}) = \sum_{\mathbf{R}_0} \int_{u.c.} d\mathbf{r} f(\mathbf{r}) e^{-i\mathbf{q}\cdot(\mathbf{R}_0+\mathbf{r})} \qquad (11.9)$$

$$= \sum_{\mathbf{R}_0} e^{-i\mathbf{q}\cdot\mathbf{R}_0} \int_{u.c.} d\mathbf{r} f(\mathbf{r}) e^{-i\mathbf{q}\cdot\mathbf{r}}$$

where the integral is now over the unit cell. We now introduce an appropriate set of crystallographic coordinates for the symmetry and recall that in these coordinates the symmetry translation vectors are expressed as $[n^j]$, i.e., a set of three integers or simple fractions. Equation (11.9) becomes:

$$F(\mathbf{q}) = \sum_{[n^j]} e^{-2\pi i q_j n^j} \int_{u.c.} d\mathbf{r} f(\mathbf{r}) e^{-i\mathbf{q}\cdot\mathbf{r}}. \qquad (11.10)$$

We have used the usual convention that contracted indices are implicitly summed. The following statement is now clear by inspecting Eqn. (11.10):

For $N \to \infty$, $F(\mathbf{q})$ *is non-zero only for* $\mathbf{q}$ *belonging to the* primitive RL.

In fact, if $\mathbf{q}$ belongs to the *primitive* reciprocal lattice, then by definition its dot product with the symmetry lattice translation is a multiple of 2π, the exponential factor is 1 and the summation yields N (i.e., the number of unit cells). Conversely, if $\mathbf{q}$ does not belong to the *primitive* reciprocal lattice, the exponential factor will vary over the unit circle in complex-number space and will average to zero. In particular, $F(\mathbf{q})$ is *zero* for the conventional RLVs that are extinct by centering (this explains the terminology "extinction").

11.5.3 Choice of the origin

It is important to stress here that the value of the Fourier transform we have just written depends on the *choice of the origin*. In fact, in order to perform the calculation, we had to write the function $f(\mathbf{r})$ in terms of a *position vector* $\mathbf{r}$, which of course corresponds to a different position in the crystal if a different choice of the origin is made. What happens to the function $F(\mathbf{q})$ for a different choice of the origin in *real* space? If the shift in the origin is such that $\mathbf{r}' = \mathbf{r} + \mathbf{r}_0$, then from the perspective of the new origin the original lattice function $f(\mathbf{r})$ will be replaced by a new lattice function $f'(\mathbf{r}') = f(\mathbf{r}' - \mathbf{r}_0)$. The Fourier transform of this new function will become:

$$F'(\mathbf{q}) = N \int_{u.c.} d\mathbf{r}' f'(\mathbf{r}') e^{-i\mathbf{r}'\cdot\mathbf{q}} = N e^{-i\mathbf{r}_0\cdot\mathbf{q}} \int_{u.c.} d\mathbf{r}' f(\mathbf{r}' - \mathbf{r}_0) e^{-i(\mathbf{r}' - \mathbf{r}_0)\cdot\mathbf{q}}$$

$$= e^{-i\mathbf{r}_0\cdot\mathbf{q}} F(\mathbf{q}) \quad (11.11)$$

where the last step in Eqn. (11.11) arises from relabeling of the integration variable. In other words:

Shifting the origin by $\mathbf{r}_0$ corresponds to multiplying each of the Fourier components by the phase factor $e^{-i\mathbf{r}_0\cdot\mathbf{q}}$.

11.5.4 The significance of the Fourier transform in crystallography

The previous statement has been deduced in a completely general way, regardless of the specific form of the function $f(\mathbf{x})$. In fact, it is the *periodic* nature of $f(\mathbf{x})$ that is responsible for the *discrete* nature of $F(\mathbf{q})$. We can now reflect on the concrete significance of the Fourier transform in the context of the theory of scattering of photons or particles from a crystal. Under certain approximations, it can be shown (see, for example, Warren, 1990) that the *scattering amplitude* for X-rays of a single crystal can be expressed as

$$A(\mathbf{q}) = r_0 \int_{\text{Crystal}} d\mathbf{R} f(\mathbf{R}) e^{-i\mathbf{q}\cdot\mathbf{R}} \left[\boldsymbol{\epsilon}\cdot\boldsymbol{\epsilon}'\right] \quad (11.12)$$

where $r_0 = e^2/4\pi\epsilon_0 mc^2 = 2.82 \times 10^{-15}$ m is known as the **classical electron radius**, $\boldsymbol{\epsilon}$ and $\boldsymbol{\epsilon}'$ are the polarization vectors of the incident and scattered photons, respectively, and $f(\mathbf{R})$ is the number density of electrons. One obtains very similar expressions for scattering of electrons, neutrons or other particles. Here, we will not dwell on the derivation of Eqn. (11.12) and on the approximations under which it is valid (briefly, the crystal must be sufficiently small compared to the interaction length of the radiation but still sufficiently large to be considered as having translational periodicity; also, the incident wavelength must not be near absorption resonances of the elements). We will instead simply remark on the similarity between Eqns (11.12) and (11.6).

11.5.5 Symmetry of the "weighed" lattice points

Given a certain periodic lattice function $f(\mathbf{r})$, typically representing the density of scattering centers, we can therefore "weigh" or "dress" each point of the RL with the value of the Fourier transform of that function or of a derived function, calculated at that particular RL point. We are mainly interested in the RL weighted with $F(\mathbf{q})$ (proportional to the scattering amplitude) and $|F(\mathbf{q})|^2$ (proportional to the cross-section) but our considerations will be completely general. The consequences of weighting the RL are quite dramatic. In fact, the RL loses one of its main properties, namely the *translational invariance*. Nevertheless, not all RL symmetry is lost. In fact, we can easily relate the values of $F(\mathbf{q})$ at RL points related by elements of the *crystal class* (a subgroup of the holohedry). To do this, we observe that, if $t \circ R$ is a symmetry operator in real space, then $f(\mathbf{t} + \mathbf{Rr}) = f(\mathbf{r})$. In addition, the infinitesimal volume element $d(\mathbf{t} + \mathbf{Rr})$ is equal to $d\mathbf{r}$, because symmetry operators are isometric. We can therefore rewrite Eqn. (11.9) as

$$F(\mathbf{q}) = N \int_{\text{u.c.}} d\mathbf{r} f(\mathbf{r}) e^{-i\mathbf{q}\cdot\mathbf{r}} \qquad (11.13)$$

$$= N \int_{\text{u.c.}} d(\mathbf{t} + \mathbf{Rr}) f(\mathbf{t} + \mathbf{Rr}) e^{-i\mathbf{q}\cdot(\mathbf{t}+\mathbf{Rr})}$$

$$= N \int_{\text{u.c.}} d(\mathbf{r}) f(\mathbf{r}) e^{-i\mathbf{q}\cdot(\mathbf{Rr})} e^{-i\mathbf{q}\cdot\mathbf{t}}$$

where we have assumed that $\mathbf{q}$ belongs to the primitive RL and that the sum is over N unit cells, N being a "very large" number. At this point, we exploit Eqn. (11.5) and rewrite

$$\mathbf{q} \cdot (\mathbf{Rr}) = q_i R^i_j x^j = (\mathbf{R}^{-1}\mathbf{q}) \cdot \mathbf{r}. \qquad (11.14)$$

With this, Eqn. (11.13) becomes

$$F(\mathbf{q}) = N \int_{\text{u.c.}} d\mathbf{r} f(\mathbf{r}) e^{-i(\mathbf{R}^{-1}\mathbf{q})\cdot\mathbf{r}} e^{-i\mathbf{q}\cdot\mathbf{t}} = F(\mathbf{R}^{-1}\mathbf{q}) e^{-i\mathbf{q}\cdot\mathbf{t}}. \qquad (11.15)$$

Equation (11.15) is the basis of a series of very important statements.

11.5.6 Centrosymmetric structures

If the structure is centrosymmetric and if the origin has been chosen to coincide with a center of inversion, then $f(\mathbf{r}) = f(-\mathbf{r})$. We can exploit this to simplify the Fourier transform significantly:

$$F(\mathbf{q}) = N \int_{\text{u.c.}} d\mathbf{r} f(\mathbf{r}) e^{-i\mathbf{q}\cdot\mathbf{r}} = N \int_{\text{u.c.}} d\mathbf{r} f(-\mathbf{r}) e^{i\mathbf{q}\cdot\mathbf{r}}$$

$$= N \int_{\text{u.c.}} d\mathbf{r} f(\mathbf{r}) e^{i\mathbf{q}\cdot\mathbf{r}} = F^*(\mathbf{q}). \qquad (11.16)$$

In other words

If the origin coincides with a center of inversion, the structure factors of a centrosymmetric structures are real.

11.5.7 Laue classes

The reciprocal lattice weighted with $|F(\mathbf{q})|^2$ has the *full point-group symmetry of the crystal class*. This is because the phase factor $e^{-i\mathbf{q}\cdot\mathbf{t}}$ in Eqn. (11.15) clearly disappears when taking the modulus squared. In fact, there is more to this symmetry when $f(\mathbf{r})$ is *real*, i.e., $f(\mathbf{r}) = f^*(\mathbf{r})$: in this case

$$F^*(\mathbf{q}) = N \int_{\text{u.c.}} d\mathbf{r} f^*(\mathbf{r}) e^{i\mathbf{q}\cdot\mathbf{r}} \qquad (11.17)$$

$$= N \int_{\text{u.c.}} d\mathbf{r} f(\mathbf{r}) e^{i\mathbf{q}\cdot\mathbf{r}} = F(-\mathbf{q}).$$

Consequently, $|F(\mathbf{q})|^2 = F(\mathbf{q}) F(-\mathbf{q}) = |F(-\mathbf{q})|^2$ is *centrosymmetric*. The lattice function used to calculate non-resonant scattering cross-sections is *real*. Therefore, the $|F(\mathbf{q})|^2$-weighted RL (proportional to the Bragg peak intensity) has the symmetry of the crystal class *augmented by the center of symmetry*. This is necessarily one of the 11 centrosymmetric point groups, and is known as the *Laue class* of the crystal.

11.5.8 The Patterson function

The Fourier transform of a lattice function $F(\mathbf{q})$ is, in the most general case, a complex number. As we have seen above, this phase information is lost in a scattering experiment, where the cross-section we measure is proportional to $|F(\mathbf{q})|^2$. Consequently, the scattering density cannot be determined without somehow "phasing" the reflections. An entire branch of crystallography is devoted to the so-called "direct" methods, which can determine the phases of a set of reflections starting from a set of "reasonable" assumptions (Warren, 1990). It is nonetheless possible to obtain some degree of information about the scattering density *without* any knowledge of the phases. Again, from Eqn. (11.13), we obtain easily:

$$|F(\mathbf{q})|^2 = N^2 \iint_{\text{unit cell}} d\mathbf{r} d\mathbf{r}' f(\mathbf{r}) f(\mathbf{r}') e^{-i\mathbf{q}\cdot(\mathbf{r}-\mathbf{r}')}. \qquad (11.18)$$

Problem 11.2: by employing a change of variable, show that the following equation holds:

$$\frac{1}{N^2 v_0} \sum_\tau |F(\tau)|^2 e^{i\tau \cdot r} = \int_{\text{unit cell}} dr' \, f(r') f(r+r') = P(r) \quad (11.19)$$

where τ are (ideally) all the RL points. The function defined in Eqn. (11.19) is known as the **Patterson function** (or "Patterson" for the cognoscenti, from Lindo Patterson, 1934). One can perhaps recognize in Eqn. (11.19) that the Patterson is **the autocorrelation function of the scattering density**.

The following are important properties of the Patterson function for atomic-like scattering densities:

- Patterson functions are three-dimensional functions defined within one unit cell, and are usually presented in the form of two-dimensional "slices".
- Atomic-like scattering densities are *mostly zero*, except at the atomic positions. Therefore the Patterson function will be mostly zero as well, except at the *origin* ($r = 0$) and *for values of r corresponding to vectors joining two atoms*. At these vectors, the Patterson function will have **peaks**.
- The height of the $r = 0$ peak can be easily calculated (left as an exercise).
- The height of the peaks for $r \neq 0$ is *proportional to the product of the scattering power of the two atoms joined by the vector r*. For example, for X-rays, the height of the Patterson peaks is $\propto Z^2$ for atoms of the same species and $\propto Z_1 Z_2$ for atoms of different species.
- If the crystal structure contains few heavy atoms and many light atoms, *one can easily see that the Patterson will be dominated by heavy-atom peaks*. The strongest family will be at r vectors joining two heavy atoms, and the next strongest the one at r vectors joining a heavy and a light atom.
- In simple cases, the Patterson can often be *uniquely interpreted*, yielding the exact position of all the atoms in the unit cell.

11.5.9 Extinction conditions due to roto-translations

If $t \circ R$ is a roto-translation (i.e., a glide plane or a screw axis) and $\mathbf{q}$ lies on the *graph* of R, then $F(\mathbf{q}) = 0$ unless $\mathbf{q} \cdot \mathbf{t}$ is an integral multiple of 2π. In fact, $\mathbf{q}$ belongs to the graph of R if $\mathbf{Rq} = \mathbf{R}^{-1}\mathbf{q} = \mathbf{q}$. Therefore, from Eqn. (11.15),

$$F(\mathbf{q}) = F(\mathbf{R}^{-1}\mathbf{q})e^{-i\mathbf{q}\cdot\mathbf{t}} = F(\mathbf{q})e^{-i\mathbf{q}\cdot\mathbf{t}}. \quad (11.20)$$

The presence of roto-translations gives rise to a series of extinctions, in addition to those due to centering. In the case of extinctions by roto-translation, the extinct RL vector $\mathbf{q}$ does belong to the primitive RL, but $F(\mathbf{q}) = |F(\mathbf{q})|^2 = 0$. In particular:

Glide plane extinctions affect RL *planes*. For example, in space group $Pnma$ (No. 62), the Hermann–Mauguin symbol contains two glide planes: $n \perp x$ and $a \perp z$. The first glide (n) affects RL vectors with fractional coordinates $[0\,k\,l]$, which are invariant by reflection through a plane perpendicular to a^*. The glide translation in fractional coordinates is $[0\,\frac{1}{2}\,\frac{1}{2}]$, so the condition for $\mathbf{q}\cdot\mathbf{t}$ to be an integral multiple of 2π is equivalent to $k+l = 2n$. The second glide (a) affects RL vectors of the form $[h\,k\,0]$. The glide translation in fractional coordinates is $[\frac{1}{2}\,0\,0]$, so the condition for $\mathbf{q}\cdot\mathbf{t}$ to be an integral multiple of 2π is equivalent to $h = 2n$.

Screw axis extinctions affect RL *lines*. For example, in space groups $P6_n$, $n=1,\ldots,5$, the screw axis affects RL vectors with fractional coordinates $[0\,0\,l]$, which are invariant under a six-fold rotation. The screw translation vectors are different for the different space groups:

- Space groups $P6_1$ and $P6_5$: the screw translation in fractional coordinates is $[0\,0\,\frac{1}{6}]$ and $[0\,0\,\frac{5}{6}]$, respectively, and, in both cases, the reflection condition is $l = 6n$.
- Space groups $P6_2$ and $P6_4$: the screw translation in fractional coordinates is $[0\,0\,\frac{1}{3}]$ and $[0\,0\,\frac{2}{3}]$, respectively, and, in both cases, the reflection condition is $l = 3n$.
- Space groups $P6_3$: the screw translation in fractional coordinates is $[0\,0\,\frac{1}{2}]$ and the reflection condition is $l = 2n$.

11.5.10 Reflection conditions in the ITC

A *reflection condition* is a condition for a particular class of vectors of the *weighted conventional* reciprocal lattice not to be extinct. For a completely generic lattice function, reflection conditions take into account extinction by both centering and roto-translation. These conditions are listed as "general" in the ITC, in a column next to the coordinates of the general positions.

11.6 Further exploitation of the RL symmetry

In deriving Eqn. (11.10), we have exploited the translational symmetry of the lattice function in real space to reduce the integration volume to one primitive or conventional unit cell. We can further reduce the integration volume to one *asymmetric* unit cell, by exploiting the additional symmetries of the lattice function. Using crystallographic coordinates explicitly

$$F(\mathbf{q}) = Nv_0 \sum_{j=1}^{M} \int_{asy.u.c.} dx^i f(x^i) e^{-2\pi i q_i x^i(j)} \qquad (11.21)$$

where M is the number of equivalent positions, $x(j)$, in the conventional unit cell. By bringing the summation inside the integral we obtain:

$$F(\mathbf{q}) = Nv_0 \int_{asy.u.c.} dx^i f(x^i) \frac{m(x)}{M} \left(\sum_{j=1}^{M} e^{-2\pi i q_i x^i(j)} \right) \qquad (11.22)$$

$$= Nv_0 \int_{asy.u.c.} dx^i f(x^i) \frac{m(x)}{M} (A + iB)$$

where

$$A = \mathfrak{Re}\left(\sum_{j=1}^{M} e^{-2\pi i q_i x^i(j)} \right) \qquad (11.23)$$

$$B = \mathfrak{Im}\left(\sum_{j=1}^{M} e^{-2\pi i q_i x^i(j)} \right).$$

Note that we have introduced the multiplicity factor $m(x)/M$, where $m(x)$ is the multiplicity of point x. This is 1 for all points in general positions, and is therefore unnecessary for a truly continuous $f(\mathbf{x})$, but is required for atomic-like lattice functions.

The functions $A(x, y, z; h, k, l)$ and $B(x, y, z; h, k, l)$ are polynomials in the sines and cosines of the real- and reciprocal-space coordinates. They do not depend on the specific form of the lattice function $f(x)$, but only on the space group. Their knowledge greatly simplifies crystallographic structure factor calculations, and are therefore listed by Shmueli (2001).

11.6.1 Special extinction conditions

It often happens that for some positions of special symmetry, both $A(x, y, z; h, k, l)$ and $B(x, y, z; h, k, l)$ are *zero* for entire classes of RL vectors. This gives rise to "special" extinction conditions that are strictly valid only if $f(\mathbf{x})$ is a delta function. It can be proven, however, that the reflections are also extinct if $f(\mathbf{x})$ is *spherically symmetric* around the special position. Fore more complex (but still atomic-like) $f(\mathbf{x})$, the meaning of the "special" extinction condition is that the "atom" at a particular special position will give no contribution to $F(\mathbf{q})$.

"Special" extinction conditions arise because some of the special positions have additional symmetries. A simple example is that of sites lying on the graph of a roto-translation operator $t \circ R$. The *rotational* component R of the roto-translation has no effect on those sites, so the *translational* component t is a symmetry operator for those particular sites *only*. This results in all the RL vectors for which $\mathbf{q} \cdot \mathbf{t} \neq 2\pi n$ to be *extinct* for those sites. Note that points lying on a roto-translation graph are in general *not* special positions, because they do not have reduced multiplicity. Therefore, "special" extinction conditions are listed only for those points on the roto-translation graph that happen to be special positions for other reasons (e.g., because they are at a center of inversion).

Example: in space group $C2/m$ (No. 12), unique axis b, standard cell choice, there are two centers of inversion in the asymmetric unit cell: $[4e\ \frac{1}{4}, \frac{1}{4}, 0]$ and $[4f\ \frac{1}{4}, \frac{1}{4}, \frac{1}{2}]$. They both lie on an a glide plane at $[x, \frac{1}{4}, z]$, which effectively *doubles* the lattice periodicity along a for points on its graph. Consequently, the "special" reflection condition $h = 2n$ applies to all RL vector classes hkl.

The majority of "special" extinction conditions are due to additional translational symmetry of the special positions, either due to glide planes and screw axes or more generally to the fact that the sublattice of the special positions with their equivalent sites has additional translational symmetry (a simple example of the latter case is found in space group $P4$, No. 75, site $2c\ 2..$). These "special" extinctions affect all RL vector classes hkl. However, instances of "special" extinction conditions affecting planes of reflections are also found quite frequently in the ITC.

The Wigner–Seitz constructions and the Brillouin zones

12

Up to this point, we have only considered the symmetry of the *RL nodes*, showing that this is in fact completely adequate when dealing with the Fourier transform of *periodic* functions. In essence, crystallography is only interested in the very sharp "spikes" of scattering at the RL nodes, ignoring completely the vast regions of reciprocal space outside these nodes. For a well-ordered crystal, scattering outside the RL is weak by comparison but by no means zero, and contains a wealth of information about *intrinsic static disorder* (elastic scattering) and *dynamics* (inelastic scattering from phonons, spin waves, etc.) General (non-RL) reciprocal-space vectors are also essential in describing phenomena such as lattice electrons, phonons, etc. In short,

Non-periodic phenomena in the crystal (elastic or inelastic) are described in terms of generic (non-RL) reciprocal-space vectors and give rise to scattering outside the RL nodes.

Although the individual excitations may break every crystal symmetry, the crystal maintains its symmetry **on average** (either temporal average or average over different regions of a large crystal). One therefore usually deals with a reciprocal space that retains the full crystal class symmetry at the very least or sometimes the Laue symmetry. In describing these phenomena, however, one encounters a problem: as one moves away from the RL origin, symmetry-related "portions" of reciprocal space will become very distant from each other. In order to take full advantage of the reciprocal-space symmetry, it is therefore advantageous to **bring symmetry-related parts of the reciprocal space together in a compact form**. This is exactly what the Wigner–Seitz constructions accomplishes very cleverly (the "Wigner–Seitz" construction is actually due to the German mathematician P.L.G. Dirichlet, 1805–1859 (Dirichlet, 1850). The term "**Brillouin zones**" is applied to the fully symmetric portions of reciprocal space, as obtained using the "extended" Wigner–Seitz construction (see below), when they are employed to

118 The Wigner–Seitz constructions and the Brillouin zones

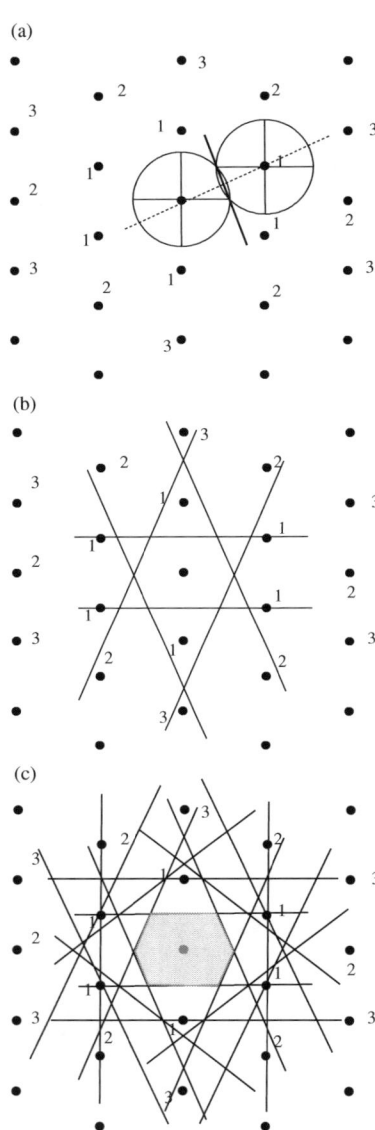

Fig. 12.1 Construction of the Wigner–Seitz unit cell for the case of a C-centered rectangular lattice in 2D. **A**: bisecting lines are drawn to the segments connecting the origin with the neighboring points (marked "1"). **B**: these lines define a polygon – the Wigner–Seitz unit cell. **C**: the Wigner–Seitz unit cell is shown together with lines bisecting segments to more distant lattice points.

describe lattice, electronic or magnetic excitations in crystals. A very good description of the Wigner–Seitz constructions and their uses in the definition of the Brillouin zone can be found in (Ashcroft and Mermin, 2003).

12.1 The Wigner–Seitz construction at the origin of reciprocal space

The Wigner–Seitz construction is essentially a method of constructing, for every Bravais lattice, a **fully symmetric** unit cell that has **the same volume as a *primitive* cell**. As such, it can be applied to both *real* and *reciprocal* spaces, but it is essentially employed only for the latter.

The Wigner–Seitz unit cell containing the origin *of reciprocal space is the set of points that are* closer *to the origin than to any other RL lattice node.*

Once defined at the origin, the Wigner–Seitz unit cell can be employed for "tiling" all of the reciprocal space with identical cells – this procedure is often named the "repeated Wigner–Seitz scheme". In the repeated scheme, for every RL node τ, the Wigner–Seitz unit cell containing τ is the set of points that are *closer* to τ than to any other node.

It is quite apparent that:

- Each Wigner–Seitz unit cell contains one and only one lattice node.
- Every point in space belongs to at least one Wigner–Seitz unit cell. Points belonging to more than one cell are *boundary* points between cells.
- From the previous two points, it is clear that the Wigner–Seitz unit cell has the same volume as a primitive unit cell. In fact, it "tiles" the whole space completely with identical cells, each containing only one lattice node.
- The Wigner–Seitz unit cell **containing the origin** (also known as the first Brillouin zone) has the full **point-group symmetry of the lattice** (holohedry). In **real space**, the origin is arbitrary, and all the Wigner–Seitz unit cells are the same. In **the "weighted" reciprocal space the Wigner–Seitz at q = 0 is** *unique* **in having the full point-group symmetry**. As we shall see shortly, the *extended* Wigner–Seitz scheme is used to project fully symmetric portions of reciprocal space away from the origin into the first Wigner–Seitz unit cell.

12.1.1 The Wigner–Seitz construction in practice (Figure 12.1)

- Draw segments connecting the origin with the neighboring points. The first "ring" of points (marked with "1" in Figure 12.1**A**) should

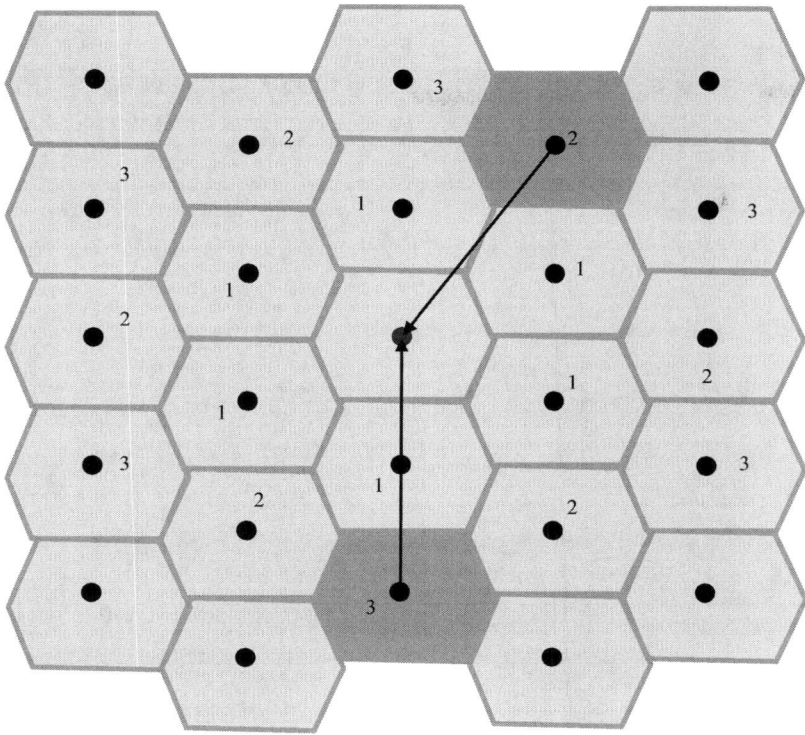

Fig. 12.2 Repeated Wigner–Seitz cell scheme, showing how the entire space can be tiled with these cells. Each cell can be "reduced" to the first Wigner–Seitz cell with a *single* RL vector.

be sufficient, although these points may not all be symmetry-equivalent.
- Draw orthogonal lines bisecting the segments you just drew (Figure 12.1**A**). These lines define a polygon containing the origin (Figure 12.1**B**) – this is the Wigner–Seitz unit cell. In 3D, one would need to draw **orthogonal bisecting planes**, yielding Wigner–Seitz **polyhedra**.
- Figure 12.1**C** shows an extended construction (to be used later) including lines bisecting the segments to the second and third "rings". As you can see, the new lines do not intersect the original Wigner–Seitz unit cell.
- The whole space can be "tiled" with Wigner–Seitz cells (Figure 12.2).

12.1.2 "Reduction" to the first Wigner–Seitz unit cell (first Brillouin zone)

As anticipated, the main use of the Wigner–Seitz unit cell is in reciprocal space: Every vector **q** in reciprocal space can be written as

$$\mathbf{q} = \mathbf{k} + \boldsymbol{\tau} \qquad (12.1)$$

where $\boldsymbol{\tau}$ is a RL vector and $\mathbf{k}$ is within the *first* Wigner–Seitz unit cell (i.e., the one containing the origin). We more often say that $\mathbf{k}$ is the "equivalent" of $\mathbf{q}$ reduced to the first Brillouin zone (see below).

The **repeated Wigner–Seitz scheme** shown in Figure 12.2 is used to determine **which $\boldsymbol{\tau}$ should be used for a given $\mathbf{q}$** – clearly, the one corresponding to the lattice node *closer* to it.

12.2 The extended Wigner–Seitz scheme: construction of higher-order Brillouin zones

We have just learned how to "reduce" every reciprocal-space point to the first Wigner–Seitz unit cell (or first Brillouin zone). But the question is: which "bits" of reciprocal space should be "reduced" together? One may be tempted to think that an entire Wigner–Seitz unit cell should be "reduced" together – after all, one would only need a *single* RL vector to accomplish this. It is readily seen, however, that this is not a good idea. As we mentioned before, higher Wigner–Seitz unit cells (i.e., other than the first) in the repeated scheme do not possess any symmetry, and we are specifically interested in "reducing" together symmetry-related parts of reciprocal space. Therefore, **a different construction, known as the extended Wigner–Seitz construction, is required to reduce symmetry-related portions of reciprocal space (higher-order Brillouin zones) simultaneously**.

The first Brillouin zone coincides with the first Wigner–Seitz unit cell. Higher Wigner–Seitz unit cells in the repeated scheme are emphatically not *Brillouin zones.*

12.2.1 The extended Wigner–Seitz scheme in practice (Figure 12.3)

- Start off in the same way as for the Wigner–Seitz construction, but with lines bisecting the segments to higher-order "rings" of points, as per Figure 12.1**C**.
- Many polygons of different shapes (polyhedra in 3D) will be obtained. Each of these will be given a **number according to how many lines (planes in 3D) are crossed to reach the origin with a straight path**. If m lines (planes) are crossed, the **order of the Brillouin zone will be** $m + 1$.
- A Brillouin zone is formed by polygons (polyhedra) having the same number (Figure 12.3**A**).

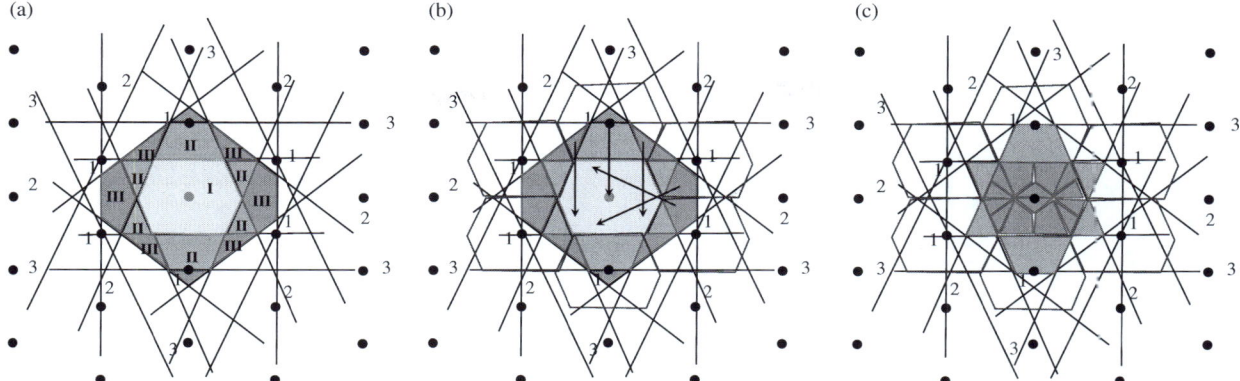

Fig. 12.3 The extended Wigner–Seitz scheme, which is employed to construct higher-order Brillouin zones. The starting point is Figure 12.1C. **A**: A number is given to each polygon, according to how many lines are crossed to reach the origin. Polygons with the same number belong to the same higher-order Brillouin zone. The figure shows the scheme for the first three Brillouin zones. **B**: Portions of a higher Brillouin zone can be reduced to the first Brillouin zone in the normal way, i.e., by using the extended and repeated Wigner–Seitz constructions (here, the reduction procedure is shown for the third zone). **C**: When reduced, higher zones "tile" perfectly within the first Brillouin zone.

- As anticipated, **the first Brillouin zone is also the first Wigner–Seitz cell** (no line is crossed).
- The different portions of a Brillouin zone are "reduced" to the first Brillouin zone in the normal way, i.e., using the extended Wigner–Seitz construction (Figure 12.3**B**).
- All the portions of a higher Brillouin zone will **tile perfectly within the first Brillouin zone** (Figure 12.3**C**).

References

Ashcroft, N.W. and Mermin, N.D. (2003). *Solid State Physics*. HRW International Editions, CBS Publishing Asia Ltd.

Barlow, W. (1884). Über die geometrischen Eigenshaften homogener starrer Strukturen. *Z. Kristallogr. Mineral*, **23**, 1–63.

Bieberbach, L. (1910). Über die bewegungsgruppen der n-dimensionalen Euklidischen Räume mit einem endlichen Fundamentalbereich. *Göttinger Nacht*, 75–84.

Bravais, A. (1850). Mémoire sur les systèmes formés par des points distribués régulièrement sur un plan or dans l'espace. *J. Ec. Polytech.*, **19**, 1–128.

Bravais, A. (1851). Etudes cristallographiques. *J. Ec. Polytech.*, **20**, 101–176.

Cotton, F.A. (1990). *Chemical Applications of Group Theory*, 3rd edn. John Wiley, New York.

Dirichlet, P.G.L. (1850). Über dir Reduktion der positiven quadratischen Formen mit drei unbestimmten ganzen Zahlen. *J. Reine Angew. Math.*, **40**, 209–227.

Fedorov, E.S. (1890). *The Symmetry of Regular Systems of Figures* (in Russian). A. Yakob, St. Petersburg, Academy of Sciences, Moscow.

Fedorov, E.S. (1891). The symmetry of regular systems of figures (in Russian). *Not. Imp. St Petersburg Mineral. Soc.*, Series 2, **28**, 1–146.

Gadolin, A.V. (1867). Deduction of all crystallographic systems and their subdivisions by means of a single general principle (in Russian). *Ann. Imp. St Petersburg Mineral. Soc. Ser. 2*, **4**, 112–200.

Giacovazzo, C., Monaco, H.L., Artioli, G., Viterbo, D., Ferraris, G., Gilli, G., Zanotti, G., and Catti, M. (2002). *Fundamentals of Crystallography*. International Union of Crystallography, Oxford University Press, New York.

Hahn, Th. (ed.) (2002). *International Tables for Crystallography*, 5th edn. Volume A. Kluwer Academic, Dodrecht.

Haüy, R.-J. (1822). *Traité de Cristallographie* (2 Volumes). Bachelier, Paris.

Hayes, H. (2011) A collection of photographs from Antakya Archaeological Museum, Antioch. http://www.sacred-destinations.com/turkey/antioch-mosaic-photos/index.html. Images by Dick Osserman.

Hermann, C.H. (1929). Zur systematischen Struktur-theorie. IV. Untergruppen. *Z. Kristallogr.*, **69**, 533–555.

Hermann, C.H. (ed.) (1935). *Internationale Tabellen zur Bestimmung von Kristallstrukturen*. Gebrü Borntrâger, Berlin.

Hermann, C.H. and Mauguin, C. (1935). In *Internationale Tabellen zur Bestimmung von Kristallstrukturen* (ed. C. Hermann), Volume 1. Gebrü Borntrâger, Berlin.

Jali (2000). Heilbrunn Timeline of Art History. The Metropolitan Museum of Art, New York. http://www.metmuseum.org/toah/works-of-art/1993.67.2 (October 2006).

Kopský, V. and Litvin, D.B. (ed.) (2002). *International Tables for Crystallography*, 1st edn. Volume E. Kluwer Academic, Dodrecht.

Lima-de Faria, J. (ed.) (1990). *Historical Atlas of Crystallography*. Kluwer Academic, Dordrecht.

Miller, W.H. (1839). *A Treatise on Crystallography*. Deighton, Cambridge.

Mohs, F. (1822). *Grundriss der Mineralogie*. Volume 1. Dresden.

Opechowski, W. and Guccione, R. (1965). Magnetic symmetry. In *Magnetism*, Vol II part A (ed. G. Rado and H. Suhl), pp. 105–165. Academic Press, New York.

Owen, J. (1865). *The Grammar of Ornaments*, Dorling Kindersley, London.

Schattschneider, D. and Hofstadter, D.R. (2004). *Visions of Symmetry: Notebooks, Periodic Drawings, and Related Work of M. C. Escher*. Thames and Hudson, London. Escher society.

Schoenflies, A. (1891). *Kristallsysteme und Kristallstruktur*. Teubner, Leipzig.

Schubnikov, A.V. and Belov, N.V. (1964). *Colored Symmetry.* Pergamon Press, Oxford.

Shmueli, U. (ed.) (2001). *International Tables for Crystallography,* 2nd edn. Volume B. Kluwer Academic, Dodrecht.

Speiser, A. (1929). *Theorie der Gruppen von endlicher Ordnung,* 2nd edn. Springer, Berlin.

Warren, B.E. (1990). *X-ray Diffraction,* 2nd edn. Dover Publications, New York.

Weiss, C.S. (1815). Uebersichtliche Darstellung der verschiedenen natürlichen Abteilungen der Kristallisation-Systeme. *Abh. K. Akad. Wiss. Berlin,* 289–344.

Weiss, C.S. (1817). Ueber eine verbesserte Methode für die Bezeichnung der verschiedenen Flächen eines Kristallisation-Systeme. *Abh. K. Akad. Wiss. Berlin,* 286–314.

Whewell, W. (1825). A general method of calculating the angles made by any planes of crystals, and the laws according to which they are formed. *Philos. Trans. R. Soc. London,* **1**, 87–130.

Wyckoff, R.W.G. (1922). *The Analytical Expression of the Results of Space Groups.* Carnegie Institute of Washington, Washington DC.

Index

Active transformation 5
Asymmetric unit 23

Basis vectors 54
Bravais lattices
  in 2D 38–40
  in 3D 83–87
Brillouin zones construction
  practical construction 120
Brillouin zones 120–121

Commutation 27
Composition
  of operators in normal form 30
Conjugation 10
Contravariant quantities 54
Coordinate system
  generalized transformation 56
Coordinate systems 51–56
  and components 53
  auxiliary Cartesian coordinates 65
  for frieze groups 30
  generalized transformation 56
  notation for 54
Covariant quantities 54
Crystal systems 14

Dual basis 62
  in 3D 64
  key formulas 64

Extinction conditions
  due to centering 103
  due to roto-translations 112
  in the ITC 113
  of special positions 114

Fourier transform 107
  and centrosymmetric structures 110
  and lattice periodicity 108
  and origin choice 109
  in crystallographic coordinates 108
  significance in crystallography 109
Frieze groups 19–34
  in the ITC 31
  symbols for 22

Glide operators
  frieze groups 20, 22

  in 3D 90
Graph symmetry 9
Graphs 4
Graphs symmetry
  relation to composition 24
Group–subgroup relations 17
Groups 6
  generators of 7
    frieze groups 24
  mathematical structure of 6
  order of 7

Hermann–Mauguin notation 12

Lattice functions
  Fourier transform of, see Fourier
    transform 107
Lattice parameters
  reciprocal 65
Lattices 36
Laue classes 111

Metric tensor 61
Metrics 61–62
Miller indices 16
Multiplication tables 8
  frieze groups 23
  graphical
    frieze groups 24
    in 3D 92
    wallpaper groups 43
  parallelogram and arrow group 8
  rectangle group 8
  rules to obtain 10
  square group 8
Multiplicity 16

Origin
  change of 56
  choice of 29

Passive transformation 5
Patterson function 111
Phase transitions
  in 2D 67–71
Planar groups, see Wallpaper groups 35
Point groups 7
  in 2D 3–17
  in the ITC 12

  and local symmetry 30
  diagrams 16
  in 3D 75–81
    derivation 78
    in the ITC 80
    non-projective 79
    notation for projective 76
    projective 76
  settings 16
  symbols 14
Points of special symmetry 7, 24
Positions
  general 16
  special 16

Reciprocal lattice 62, 102
  holohedry of 105
  symmetry of 110
Reciprocal space 62, 100–115
  vs. real space – recapitulation 101
  orientation of 63
Reference frame 14
Reflection conditions, see Extinction
  conditions 113
Reflection operators
  frieze groups 19
  point groups 3
Restriction 36–37
Rotation matrices 58
Rotation operators
  compositions with mirrors and
    glides 42
  frieze groups 19
  generalized in 3D 75
  improper 90
  point groups 3
Roto-translations
  in 3D 89

Schoenflies notation 77
Screw axes 89
Space groups 89–97
  construction 93–97
Structure factor polynomials 113
Subgroups 7
Symmetry
  around a fixed point 3
  graph 11
  of an arrow 3

of a parallelogram 3
 of a rectangle 4
 of a snowflake 12
 of a square 4
 site 16
Symmetry directions 13
Symmetry operators 5
 application of 6
 composition of 6, 9
  in 2D 41
 graphical notation in 3D 91
 graphs of 6
  frieze groups 22

 in Cartesian coordinates 57
 in generalized coordinates 59
 mathematical form of 57–60
 normal form 28
 vs. coordinate transformations 60

Translation operators
 composition mirrors and glides 42
 composition with rotation axes 41
 frieze groups 19, 22
 symmetry of set 36
 transformation by graph symmetry 35

Unit cell 23
 in 2D 40

Wallpaper groups 35–50
 analyzing artwork with 45
 construction from symbols 49
 description of the 17, 45
 nomenclature 43
Wigner–Seitz cell 118–120
 practical construction 118
 reduction to first zone 119
Wyckoff letters 16

Printed and bound by CPI Group (UK) Ltd, Croydon, CR0 4YY